Augmented Humanity

Peter T. Bryant

Augmented Humanity

Being and Remaining Agentic in a Digitalized World

Peter T. Bryant
IE Business School
IE University
Madrid, Spain

ISBN 978-3-030-76444-9   ISBN 978-3-030-76445-6   (eBook)
https://doi.org/10.1007/978-3-030-76445-6

Cover illustration: © Alex Linch/shutterstock.com

This Palgrave Macmillan imprint is published by the registered company Springer Nature Switzerland AG.
The registered company address is: Gewerbestrasse 11, 6330 Cham, Switzerland

Preface

Digitalization is transforming the contemporary world, as digital technologies infuse all domains of human experience and striving. The effects are everywhere, profound and accelerating. Among the most remarkable effects is the growing capability of human-machine interaction, which allows advanced artificial agents to collaborate with human beings, as enablers, partners, and confidantes. Increasingly, human and artificial agents work together in close collaboration, effectively as one agent in many situations, to pursue common goals and purposes. In this way, digitalization is driving the augmentation of humanity, where humanity is conceived as communities of purposive, goal-directed agents. The transformation is already visible in numerous expert domains, in which artificial agents extend and complement human expertise, for example, in clinical medicine and the piloting of aircraft. Much practice in these domains now relies on the real-time assistance of humans by artificial agents, who together form digitally augmented agents, also called human-agent systems in computer science. Clinicians collaborate with such agents to perform advanced diagnosis, patient monitoring, and surgery, while pilots work closely with artificial avionic agents to control their aircraft. Within each domain, collaboration between human and artificial agents increases the accuracy, speed, and efficiency of action, although digital innovations also bring new risks and dilemmas. When collaboration fails, the results can be debilitating, costly, and sometimes fatal.

Similar opportunities are emerging in many other domains. Highly intelligent, artificial agents are becoming ubiquitous throughout human experience, thought, and action. Industrial organizations are very clearly impacted. Intelligent robots and artificial agents already perform many complex manufacturing, logistical, and administrative tasks, and some are now capable of advanced analytical and creative work (Ventura, 2019). They are powering the fourth industrial revolution. Comparable innovations are transforming education and entertainment, especially in online environments. Everyday life is equally affected. People constantly collaborate with artificial agents, such as Apple's Siri and Amazon's Alexa, in shopping, managing the home environment, and searching for information. In like fashion, the expression of personality, identity, and sociality are mediated by smartphones. Human beings curate their virtual selves, relationships, and social world in collaboration with artificial agents. Most people can confirm this from personal experience. In summary, the early twenty-first century is witnessing the rapid digital augmentation of humanity.

Furthermore, the next wave of digital technology promises to be even more transformative and collaborative. Many people could be surprised by the nature and speed of innovation. For example, newer digital technologies will empower empathic relating, allowing artificial agents to interpret and exhibit emotional states and moods. Google's recent experiments provide strong evidence: observers could not tell the difference between a human and artificial agent, during an everyday telephone conversation (Leviathan & Matias, 2018). The artificial agent sounded genuinely empathic and human. Related innovations will imitate other aspects of personality, including the expression of attitude, opinion, and humor. The cumulative impact will approach what some refer to as the singularity, in which artificial intelligence is functionally equivalent to human and possibly becomes transcendent (Eden et al., 2015). In addition, a vast array of intelligent sensors and the internet of things will enable real-time, precise perception of the environment. Ubiquitous and potentially invasive, digitally augmented surveillance will envelop the world. Applications will deliver many benefits, including the control of autonomous vehicles and smart cities (Riaz et al., 2018), and hopefully more sustainable management of the natural and built environments. Indeed, if wisely employed, digital innovations could help to alleviate the existential threats currently faced by humanity, including climate change, pandemic disease, and environmental degradation.

Many of these digital innovations rely on the fusion of multiple disciplines and particularly computer science, branches of engineering, social and cognitive psychology, and neuroscience. Major research efforts are underway which connect these fields, for example, in the study of brain-computer interfaces and multi-agent systems. Applications are already deployed in intelligent prosthetics, automated transport systems, and the hands-free use of computers (Vilela & Hochberg, 2020). Other innovations are transforming rehabilitation after brain injury and intersubjective communication (Brandman et al., 2017; Jiang et al., 2018). Overall, future digital technologies will be faster, more powerful, accurate, and connected. They will enable radically new forms of human-machine interaction and collaboration. Artificial agents will augment the full range of human thought, feeling, and action, at every level of personal and collective organization. The digital augmentation of humanity is underway.

Yet, as noted earlier, there are significant risks and dilemmas too. First, the potential benefits of digitalization are unevenly distributed. Major segments of humanity are being left behind or marginalized (The World Bank, 2016). For these groups, the digital divide is widening. If this continues, digitalization will amplify, rather than mitigate, socioeconomic deprivation, inequality, and injustice. Second, digitalization is vulnerable to manipulation by powerful interests. Owing to the technical reach and integration of digital networks, they could become tools of social control and coercion. Indeed, we already see examples of artificial systems being used to suppress, mislead, or demonize groups for ideological, political, and cultural reasons. Third, social biases and racial stereotypes easily infect artificial intelligence and machine learning, leading to new forms of discrimination and injustice. Fourth, human agents are often slow to learn and adapt, especially with respect to their fundamental beliefs and commitments. At every level, whether as individuals, groups, or collectives, people get stuck in their ways. Habits and routines are often hard to shift, and assumptions are resilient, which can be appropriate and welcome in some contexts but constraining in others. In any case, when it comes to information processing, humans are simply no match for artificial agents. Hence, humanity could struggle to absorb the effects of digitalization. Artificial agents might outrun or overwhelm the human capability for adaptive learning, resulting in unintended consequences and dysfunctional outcomes.

New Problematics

Collectively, these trends pose new explanatory problematics, defined as fundamental and often contentious questions within a field of enquiry (Alvesson & Sandberg, 2011). In this case, the field of enquiry concerns the explanation of civilized humanity, conceived as communities of purposive, goal-directed agents. The new problematics which digital augmentation poses go to the heart of this field: how can human beings collaborate closely with artificial agents while remaining genuinely autonomous in reasoning, belief, and choice; relatedly, how can humans integrate digital augmentation into their subjective and intersubjective lives while preserving personal values, identity, commitments, and psychosocial coherence; how can digitally augmented institutions and organizations, conceived as collective agents, fully exploit artificial capabilities while avoiding extremes of digitalized docility, dependence, and domination; how can humanity ensure fair access to the benefits of digital augmentation and not allow them to perpetuate systemic deprivation, discrimination, and injustice; and finally, the most novel and controversial challenge, which is how will human and artificial agents learn to understand, trust, and respect each other, despite their different levels of capability and potentiality. This last question is controversial because it implies that artificial agents will exhibit judgment and empathy. It suggests that soon we will attribute a type of autonomous mind to artificial agents. Recent research in artificial intelligence and machine learning shows that these qualities are feasible and within reach (Mehta et al., 2019).

In fact, modernity has puzzled over similar questions since the European Enlightenment. Scholars have long debated the limits of human capability and potentiality, and how best to liberate autonomous reason and choice, in light of natural and social constraints (Pinker,

2018). Modernity therefore focuses on similar questions to those posed by digitalization. These modern problematics include: how can human agents collaborate with each other, in collective thought and action, while developing as autonomous persons; how can humans absorb change while preserving valued commitments and psychosocial coherence; how can societies develop effective institutions while avoiding excessive docility and domination of citizens; how can humanity ensure fair access to the benefits of modernity and not allow growth to perpetuate deprivation, discrimination, and injustice; and finally, a defining question for modernity, which asks to what degree can and should human beings overcome their natural and acquired limits, to be more fully rational and empathic? This last question is also controversial. Some argue that limited capabilities and functional incompleteness are humanizing qualities, while others view them as flaws which can and should be overcome (Sen, 2009).

Therefore, digitalization leads us to problematize the underlying assumptions and concerns of modernity. First, digitalization radically expands intelligent processing capabilities, thus problematizing concepts of bounded rationality. Second, and relatedly, as intelligent machines edge closer to exhibiting a type of autonomous mind, they problematize the classic ontological distinction between human consciousness and material nature, and thus between mind and body. Third, digitalization supports the rapid composition and recomposition of agentic forms, allowing for dynamic modalities, thereby problematizing the traditional distinction between individuals, groups, and larger collectives. Fourth, digitalization enables adaptive commitments across multiple contexts and cultures, akin to digitally augmented cosmopolitanism. Each problematization has practical implications as well because people think and act based on their core assumptions and commitments. If these are disrupted or refuted, people must respond. Evidence suggests that some will simply resist and remain wedded to priors, whereas others may feel overwhelmed by digitalization and perhaps abandon priors altogether and surrender to digital determination. None of these extremes is an effective response. Finding the right balance will be critical. Much of this book is about that challenge.

About This Book

The book began by exploring digitalization through a set of smaller projects, each focused on a specific area of impact, especially regarding theories of problem-solving and organization. Some of the topic chapters reflect these origins. However, this approach proved too incremental and encumbered. Constraining assumptions were everywhere, from bounded rationality to the polarization of mind and nature, individuality versus collectivity, and interpretative understanding versus causal explanation, plus related distinctions between abstract and practical reason, ideal versus actual performance, and deductive versus inductive justification of belief. In summary, after a few years of struggling within such constraints, I decided that a different kind of project was required. The current book began to take shape.

Over time, I realized that the conceptual architecture of modernity was inadequate for the task. The effects of digitalization are too deep and novel, and especially the dilemmas which arise from the combination of human and artificial capabilities and potentialities. In this regard, humans are holistic in their thinking, often myopic or nearsighted, relatively sluggish in processing, and insensitive to variance. By comparison, artificial agents are increasingly focused, farsighted, fast, and hypersensitive to minor fluctuations. Clearly, human and artificial agents possess different processing capabilities and potentialities. When they collaborate as augmented agents, therefore, the result could be extremely divergent or convergent processing. One agent might dominate the other and the overall system will be convergent in human or artificial terms. Alternatively, the two types of agents might collaborate but also diverge and conflict. The combined system could be farsighted and nearsighted, fast and slow, complicated and simplified, hypersensitive and insensitive, all at the same time. However, these novel dynamics are difficult to capture using the traditional concepts and questions of modern human science. Indeed, as I argued previously, digitalization leads us to problematize many traditional assumptions and concerns. I therefore introduce fresh concepts and terminology to describe these novel phenomena, their related mechanisms, and dilemmas.

For some readers, the new concepts and terminology may be challenging. Such novelties require effortful reading, especially when used in original theorizing. However, I hope readers will agree that the uniqueness of the phenomena warrants this approach, and that to unpack and analyze the digital augmentation of humanity, we need to refresh our conceptual architecture. Significant features of digital augmentation are highly novel and not yet clearly conceptualized in the human sciences. For similar reasons, this work is largely conceptual and prospective. It looks forward, trying to shed light on an emerging terrain. It explores problematics and invites scholarly reflection about widely assumed concepts and models. Further empirical investigations and testing are necessary, of course, but first, the theoretical framework can be established.

Opportunities and Risks

This book therefore examines the prospects for digitally augmented humanity. It problematizes prior assumptions and formulates new questions and dilemmas, although the book does not attempt fully to resolve these issues. Rather, it seeks to advance the future science of augmented humanity and agency. Reflecting this breadth and style, the book is multidisciplinary, prospective, and occasionally speculative, which has advantages and risks. In terms of advantages, prospective theorizing can bring clarity and organization to new phenomena. Furthermore, it allows us to combine insights from different fields, in ways which transcend existing knowledge and which cannot easily be demonstrated empirically. More specifically, the argument combines insights from social cognitive psychology, computer science and artificial intelligence, behavioral theories of problem-solving, theories of social choice and microeconomics, and insights from organization theory, philosophy, and history. Extensive referencing of literature supports this breadth. Overall, therefore, the book looks forward toward a larger project of investigating, explaining, and managing the phenomena in question while acknowledging that initial proposals will need to evolve, as future investigations unfold.

In terms of risks, prospective theorizing is exactly this, prospective and not yet fully elaborated or tested. Moreover, problematization is inherently broad, at a high level, and hence this work does not delve into detail on every topic. This adds to the challenge and risk, but also the opportunity, for it allows us to formulate new explanatory frameworks. To do this, the author must be diligent and informed about main features of the fields in question, and the reader should be happy to think about broad questions. In further defense of this approach, the book's central motivation deserves repeating, namely, that the speed, novelty, and impact of digital augmentation require new concepts and theorizing. Extraordinary digital innovations are rushing ahead, and exploratory leaps are required to keep up. A piecemeal treatment will not do justice to the full impact of this transformation. The phenomenon calls for creative thinking at an architectural level. It calls for pluralistic "theory-driven cumulative science" (Fiedler, 2017, p. 46). My work embraces this challenge and keeps it in front of mind. As always, the reader will judge if the potential advantages outweigh and justify the risks.

Reading This Book

As noted previously, this book views humanity in terms of purposive agency and then examines how digitalization enables augmented agentic form and functioning, conceived as close human-machine collaboration. Chapters examine implications for a range of domains, from problem-solving to the future of human science. Therefore, the book is broad in scope and intended for a wide readership, embracing all the human and digital sciences. For this reason, some readers may find parts of the argument unfamiliar and technical. However, no formal or unusual methods are employed, and I expect all readers can understand what the book seeks to convey. The figures and diagrams aim to clarify the processes of digital augmentation. Each is accompanied by a full narrative explanation as well. These elements are fundamental to the work and sprinkled throughout. I therefore encourage readers to embrace the purpose, be ready to adopt new concepts and terms, and to study the figures which illustrate novel processes and mechanisms. Hopefully, readers will agree that the effort is worthwhile and that the argument lays the groundwork for further research into these phenomena.

Regarding specific chapters, it is important to start with the first two. Chapter 1 sets out the broad topic and discusses how to model the digital augmentation of humanity. Chapter 2 then identifies major historical patterns and highlights the role of technological innovation in assisting agentic capability and potentiality and, critically, the role of digital technologies in this regard. Subsequent chapters examine the implications of digital augmentation for the following domains: agentic modality, problem-solving, empathy with other minds, self-regulation, evaluation of performance, learning, self-generation, and finally the science of digitally augmented agency.

Madrid, Spain   Peter T. Bryant

References

Alvesson, M., & Sandberg, J. (2011). Generating research questions through problematization. *Academy of Management Review, 36*(2), 247–271.

Brandman, D. M., Cash, S. S., & Hochberg, L. R. (2017). Human intracortical recording and neural decoding for brain-computer interfaces. *IEEE Transactions on Neural Systems and Rehabilitation Engineering, 25*(10), 1687–1696.

Eden, A. H., Moor, J. H., Søraker, J. H., & Steinhart, E. (2015). *Singularity hypotheses*. Springer.

Fiedler, K. (2017). What constitutes strong psychological science? The (neglected) role of diagnosticity and a priori theorizing. *Perspectives on Psychological Science, 12*(1), 46–61.

Jiang, L., Stocco, A., Losey, D. M., Abernethy, J. A., Prat, C. S., & Rao, R. P. N. (2018). Brainnet: A multi-person brain-to-brain interface for direct collaboration between brains. *arXiv preprint arXiv*:1809.08632.

Leviathan, Y., & Matias, Y. (2018). Google duplex: An AI system for accomplishing real-world tasks over the phone. https://research.google/pubs/pub49194/

Mehta, Y., Majumder, N., Gelbukh, A., & Cambria, E. (2019). Recent trends in deep learning based personality detection. *Artificial Intelligence Review*, 1–27.

Pinker, S. (2018). *Enlightenment now: The case for reason, science, humanism, and progress*. Penguin.

Riaz, F., Jabbar, S., Sajid, M., Ahmad, M., Naseer, K., & Ali, N. (2018). A collision avoidance scheme for autonomous vehicles inspired by human social norms. *Computers & Electrical Engineering, 69*, 690–704.

Sen, A. (2009). *The idea of justice*. Harvard University Press.

Ventura, D. (2019). Autonomous intentionality in computationally creative systems. In *Computational creativity* (pp. 49–69). Springer.

Vilela, M., & Hochberg, L. R. (2020). Applications of brain-computer interfaces to the control of robotic and prosthetic arms. In *Handbook of clinical neurology* (Vol. 168, pp. 87–99). Elsevier.

World Bank. (2016). *World development report 2016: Digital dividends*. The World Bank.

Acknowledgments

In writing this book, I received much support and encouragement. The project began while I was employed at IE University in Madrid, Spain. Being an innovative and entrepreneurial institution, IE provided a supportive and conducive environment. I also hold a research fund there which supported the work, including the welcome opportunity to publish it open access. Throughout the process, I received invaluable advice and feedback from colleagues and friends at my own school and elsewhere. Special thanks to Richard Bryant, Pablo Garrido, Dan Lerner, Luigi Marengo, Willie Ocasio, Simone Santoni, and especially to Giancarlo Pastor who provided exceptional support and insight during a difficult year of global pandemic, when much of this book was written. Sincere thanks also to colleagues at academic conferences over recent years and to the very professional editorial and production teams at Palgrave Macmillan.

Contents

List of Figures

List of Tables

1

Modeling Augmented Humanity

As intelligent sociable agents, human beings think and act in autonomous and collaborative ways, finding fulfillment in communities of shared meaning and purpose. Without such qualities, culture and civilization would be impoverished, in fact, barely possible. Thus conceived, being and remaining agentic matter greatly. Individuals, groups, and collectives dedicate significant effort and resources to furthering these ends. Technologies of various kinds often assist them. Many institutions also exist for these purposes: to foster human development, facilitate cooperation, and grow collective endowments. Societies therefore organize to sustain and develop their members. Agentic capabilities and potentialities improve and humanity can prosper. Over recent centuries, especially, this has led to major advances in health, education, productivity, and empowerment.

Yet positive outcomes are not guaranteed. History teaches that natural and human disasters are never far away and often lead to unreasonable and inhumane behavior. Indeed, global threats loom today, including climate change, pandemic disease, and degradation of the environment. In addition, human malice and injustice can occur anywhere at any time, and they often do. Furthermore, resources and opportunities remain

© The Author(s) 2021
P. T. Bryant, *Augmented Humanity*, https://doi.org/10.1007/978-3-030-76445-6_1

scarce for many, severely restricting their potential to develop and flourish. At the same time, capability and potentiality are unevenly distributed. Humans are limited by nature and nurture, especially regarding the capabilities required for intelligent thought and action. People gather and process information imperfectly, often in myopic or biased ways, then reason and act poorly, falling short of preferred outcomes and failing to learn. Not surprisingly, therefore, it takes time and effort to grow capabilities and potentialities, especially for purposive, goal-directed action. It is the work of a lifetime, to be fulfilled as an autonomous, intelligent, and efficacious human being. And the work of history, to achieve such fulfillment on a social scale.

Historical Patterns

Notwithstanding these challenges, capabilities and potentialities develop over time, owing to improved nurture, resources, opportunities, and learning. Major drivers also include social and technological innovation (Lenski, 2015). In fact, since the earliest periods of civilization, humans have crafted tools to complement their natural capabilities. They also pondered the stars and seasons and developed explanatory models which made sense of the world and life within it, where models, in this context, are defined as simplified representations of states or processes, showing their core components and relations (see Johnson-Laird, 2010; Simon, 1979). Granted, in the premodern period, models of the world and being human often relied on myth and superstition, but they captured broad patterns nonetheless and codified the rhythms of nature and fortunes of fate. Where, following others, I define premodern as before the modern period of European Enlightenment and industrialization (e.g., Crone, 2015; Smith, 2008). And importantly, the technological assistance of humanity began in premodernity, albeit in a primitive fashion. Over time, capabilities and technologies continued to evolve and diffuse. Despite episodic disruption and setbacks, civilized humanity has developed, typically in a path-dependent fashion (Castaldi & Dosi, 2006). For Western civilization, this path traces back to ancient Greece and Rome, which in turn drew deeply from earlier, Eastern civilizations. Their cumulative legacy survives today, in many of the languages, concepts, and models which still enrich culture and thought.

Ancient learning enjoyed a renaissance in parts of the Mediterranean world during the fifteenth century CE. Artists, scholars, and architects drew insight and inspiration from the ancients. Another important inflection point was the European Enlightenment of the seventeenth and eighteenth centuries. Over time, intelligent capabilities grew, initially among privileged members of society. After much historic struggle and social change, these capabilities diffused and deepened, to become the shared endowment of modernity. Here again, technological innovation was crucial. From the first telescopes and microscopes to the printing press and early adding machines, then to the steam, electronic, and computer ages, technological innovation has expanded agentic capability and potentiality. Adam Smith (1950, p. 17) noted this type of impact, when he wrote, "the invention of a great number of machines which facilitate and abridge labour, and enable one man to do the work of many." In parallel, new psychological and social models emerged, which assume that human beings have the potential to learn and develop as intelligent agents (Pinker, 2018). The modern challenge thus became, how to grow agentic capabilities and potentialities, so that more persons can enjoy these benefits and flourish. Political and cultural struggles also ensued, as groups fought to control the future and either to defend or to dismantle the vestiges of premodernity.

While the preceding historical account is reasonably grounded, it clearly simplifies. Almost by definition, periods of civilization span widely in time and culture. Any detailed history will be notoriously complex and irregular. There are few consistent patterns, and even those which can be observed should be treated as contingent (Geertz, 2001). For the same reasons, totalizing conceptions often over-simplify. As Bruno Latour (2017) explains, modern concepts of the globe and humanity itself assume unified categories which obscure fundamental distinctions. Hence, we must ask, is it possible to identify patterns of civilized humanity over time? Previous attempts have often been misguided and lacked validity. Most transparently failed, because they sought to generalize from one or other historical context, and then extrapolated from temporal contingency to universality. Arguably, the model of history offered by

Karl Marx exhibits this flaw. Noting this common failing, it can be argued that all knowledge of such phenomena is contextual. Few, if any, patterns transcend historical contingency. To assume otherwise could be misleading and potentially dangerous, especially if it supports ideologies which deny the inherent diversity of human aspiration and experience. Nevertheless, if we respect caution and openly acknowledge contextual contingency, it is still possible to generalize, at least at a high level.

Given these caveats, scholars observe that civilized humanity exhibits broad patterns of behavior and striving over successive historical periods (Bandura, 2007). Many of these patterns are anthropological and ecological, rather than historical, in a detailed narrative sense. Evidence shows that civilized humanity has always been purposive and self-generating, creative and inventive, hierarchical and communal, settled as well as exploratory, competitive and cooperative. In short, civilized humanity is deeply agentic. Granted, these patterns are broad, but they are consistent, nonetheless. Scholars in numerous fields recognize them (e.g., Braudel & Mayne, 1995; Markus & Kitayama, 2010; Wilson, 2012). In any case, like all theoretical modeling, it is necessary to simplify, to focus on the main topics of interest. All models and theories must be selective. Debate is then about what to select and simplify, how, and why. Whether such models are illuminating and explanatory is determined by application and testing. Science always progresses in this fashion. The current work will focus on the broad effects of digital augmentation on humanity, viewed as cultural communities of purposive agents.

Dilemmas of Technological Assistance

Technologies assist and complement human capabilities, compensating for weaknesses and helping to overcoming limits. More specifically, technological assistance addresses the following needs. First, humans are limited by their physiological dependencies, whereas technologies can function independent of such constraints, for example, by operating in extreme, hostile environments. Second, humans are frequently proximal and nearsighted, whereas technologies are distal and farsighted.

Technologies therefore extend the range and scope of functioning, as when telescopes gather information from distant galaxies. Third, humans are often relatively slow and sluggish, compared to technologies which can be fast and hyperactive. Technologies therefore accelerate functioning. Fourth, humans are frequently insensitive to variance and detail, whereas technologies can be very precise and hypersensitive. In this fashion, technologies improve the accuracy and detail of functioning. Fifth, humans are irregular bundles of sensory, emotional, and cognitive functions, whereas most technologies are highly focused and coordinated. Hence, technologies enhance the reliability and accuracy of specific functions, for example, in robot-controlled manufacturing. And sixth, humans are distinguished as separate persons, groups, and collectives, while technologies can be tightly compressed, without significant boundaries or layers between them. Technologies thereby enhance functional coordination and control, exemplified by automated warehouses and factories.

All six types of extension reflect the fundamental combinatorics of technologically assisted humanity, that is, the combination of human and technological capabilities in agentic functioning. Over longer periods, history exhibits a process of punctuated equilibrium in these respects. During these punctuations, the technological assistance of agency achieves significantly greater scale, speed, and sophistication. Not surprisingly, transformations of this kind are consistent foci of study (Spar, 2020). For instance, studies investigate how modern mechanization combines technologies and humans in social and economic activity (Leonardi & Barley, 2010). Early sites were in cotton mills and steam-powered railways. Other technologies infused social and domestic life, combining humans and machines in systems of communication and entertainment. More recently, human-machine combinatorics reach into everyday thought and action, through smartphones, digital assistants, and the ubiquitous internet. Once again, the technological assistance of agency is transitioning to a new level capability and potentiality. Major benefits include far greater productivity and connectivity.

Digitalization therefore continues the historic narrative of modernity, for good and ill, where digitalization is defined as the transformation of goal-directed processes which lead to action—that is, the transformation of agentic processes—through the application of digital technologies (Bandura, 2006). Thus defined, digitalization embraces a wide range of digital technologies and affects a wide range of agentic modalities and functional domains. Most notably, advanced digital technologies enable close collaboration between human and artificial agents as digitally augmented agents, also known as human-agent systems in computer science. New challenges thus emerge. On the one hand, advanced artificial agents are increasingly farsighted, fast, compressed, and sensitive to variation. On the other hand, humans are comparatively nearsighted, sluggish, layered, and insensitive to variance. Clearly, both agents possess complementary but different capabilities, and combining them will not be easy.

Digitalization therefore entails new opportunities, risks, and dilemmas for human-machine collaboration. One possible scenario is that artificial agents will overwhelm human beings and dominate their collaboration. The overall system would be convergent in artificial terms. Alternatively, persistent human myopia and bias could infect artificial agents, and digitalization would then amplify human limitations. Now the system would be convergent in human terms. While in other situations, both types of agent may lack appropriate supervision and go to divergent extremes, where supervision in this context means to observe and monitor, then direct a process or action. In fact, we already see evidence of each type of distortion. Digitalization therefore constitutes a historic shift in agentic capability, potentiality, and risk. As in earlier periods of technological transformation, humanity will need new methods to supervise human-machine collaboration in a digitally augmented world. Analysis of these developments is a major purpose of this book. Also for this reason, it is important to distinguish the following types of agency which are central to the argument:

1. **Human agents** are organic actors. Their processes are not inherently technological, but primarily neurophysiological, behavioral, and social. Thought and action are directed by biological and psychosocial processes. At most, technologies play a supportive role in human agency.
2. **Artificial agents** are technological actors. Their processes are not inherently biological or organic, but rely on digital systems. Performance is directed by artificial, intelligent processes. The most advanced agents of this kind are fully self-generating and self-supervising. At most, humans play a supportive role in artificial agency.
3. **Augmented agents** are close collaborations between human and artificial agents. Hence, within augmented agency, processes are both

digitalized and humanized, to significant degrees. Performances are collaborative achievements of human and artificial agents.

Supervisory Challenges

Figures 1.1 and 1.2 illustrate the supervisory challenges just described, namely, how to combine and manage human and technological functioning in agentic action. The two figures depict complementary models. Both show the complexity of human functioning on the vertical axis and of technological functioning on the horizontal axis. The models also show the limits of overall supervisory capabilities, depicted by the curved lines, with L_1 being the natural human baseline, and increasing capabilities in L_2 and L_3 owing to technological assistance. In each model, the gap between these lines therefore depicts the variance in supervisory

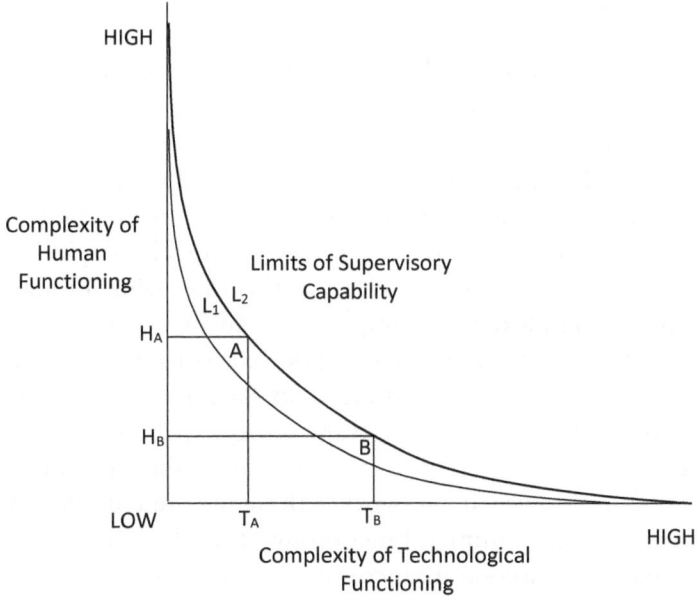

Fig. 1.1 Minor gap in supervisory capability

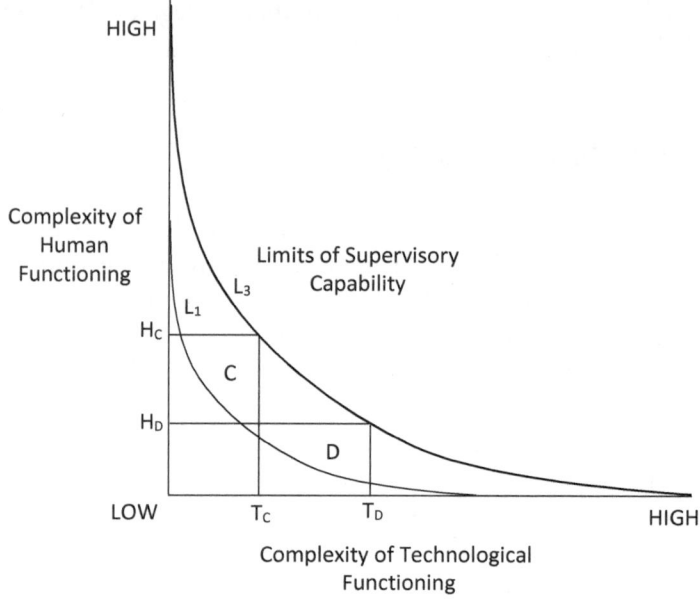

Fig. 1.2 Major gap in supervisory capability

capabilities, while all capabilities reach asymptotes of maximum complexity for human and technological functioning.

Figure 1.1 depicts a minor gap between supervisory capabilities L_1 and L_2, meaning that technological assistance at L_2 does not add much to the baseline at L_1. The figure then depicts two systems of functioning. The first is defined by human functioning H_A and technological functioning T_A. Technological complexity is less in this case, while human functioning is more complex. For example, it could be a deliberate intentional type of action which is modestly supported by technology, such as writing a letter using a pen. Assuming L_1 as the natural human baseline, an agent requires a small increase in supervisory capabilities to complete this activity, and hence capabilities at level L_2 are sufficient. Put simply, pens are simple tools, even if the written thoughts are complex. The second system is defined by human functioning H_B and technological functioning T_B. Now technological functioning is more complex, such as routine procedures which rely heavily on technologies, for example, riding in a carriage. Indeed, most people easily ride as passengers in carriages, although the carriage itself requires active supervision to maintain and control it. Once again, agents require modest supervision to complete this activity, and hence capabilities at level L_2 are sufficient. In addition, Fig. 1.1 shows two segments labeled A and B, which are the functions beyond baseline supervisory capability L_1. Both segments are relatively small, owing to the modest increase in supervisory complexity between L_1 and L_2. Put simply, it is relatively easy to supervise the use of pens and riding in carriages. In fact, many premodern activity systems were like this, owing to the relative simplicity of their technologies.

Next, Fig. 1.2 depicts a major gap between limits L_1 and L_3, meaning that technological assistance at L_3 adds significantly to the baseline at L_1, especially if L_3 includes digital technologies. The model again depicts two systems of functioning. The first is defined by more complex human functioning H_C and less complex technological functioning T_C. For example, it could be a deliberate, intentional form of action which is supported by digital technology. Perhaps the writer now uses a word processor to compose a news article. The tool may be fairly easy to use, while the thoughts are intellectually complex. Hence, the activity requires a greater level of overall supervisory capability at level L_3. Furthermore, segment C in model in Fig. 1.2 is much larger than A in model Fig. 1.1. This means that more functionality lies beyond baseline capabilities, and the overall system requires more sophisticated supervision, which is true for writing using a computer, compared to using a pen.

The second system in the model in Fig. 1.2 is defined by less complex human functioning H_D and more complex technological functioning T_D. For example, it could be a routine activity which is automated by advanced digital technology. Perhaps the passenger now rides in an autonomous vehicle, rather than sitting in a carriage. In fact, we could map the spectrum of mobility systems along the horizontal axis, from less complex systems to the most advanced artificial agents. In all cases, the overall activity system requires greater supervisory capabilities at L_3. In addition, segment D is much larger than B in Fig. 1.1. Far more functionality lies beyond baseline capabilities. The supervisory challenges are high. In terms of the example just given, it requires a significant advance in capabilities to supervise human engagement with autonomous vehicles.

Given these examples, Fig. 1.2 illustrates the supervisory challenge in highly digitalized contexts. Segments C and D show the scale of the challenge, as significant functions are beyond baseline supervisory capabilities. These segments also illustrate the functional losses which may occur if supervisory capabilities fall below level L_3. Put simply, inadequate supervision will lead to poorly coordinated action. For example, technological processes might outrun and overtake human inputs, and thereby relegate humans to a minor role in some activity. Automated vehicles could override or ignore human wishes. Alternatively, human processes may import myopias and biases, and artificial agents then reinforce and amplify human limitations. Perfectly written news articles can be racially biased and discriminatory. In both scenarios, poor supervision skews collaborative processing and leads to functional losses. Strong collaborative supervision will therefore be required, involving human and artificial agents, to ensure that both types of agent work effectively together with mutual empathy and trust.

Period of Digitalization

Digitalization therefore continues the historical narrative of technologically assisted human agency. Moreover, advanced digital systems are intelligent, self-generative agents in their own right (Norvig & Russell, 2010). Where self-generation in this context means to produce or reproduce oneself without external guidance and support. Like human beings, artificial agents are situated in the world, sensory, perceptive, calculating, and self-regulating in goal pursuit. Artificial agents also gather and process information to identify and solve problems, thereby generating knowledge and action plans. Also like humans, artificial agents are autonomous to variable degrees. In fact, the most advanced artificial agents are fully self-generating and self-supervising, meaning they generate and supervise themselves without external guidance or support. Finally, artificial and human agents are equally connected in collaborative relationships and networks.

Given these developments, human and artificial agents increasingly collaborate with each other as augmented agents. Digitalization connects them, and their combinatorics are deepening. Human and artificial agents are becoming jointly agentic, at behavioral, organizational, and even neurological levels (Kozma et al., 2018; Murray et al., 2020). So much so, that artificial and human agents will soon be indistinguishable in significant ways, approaching what some refer to as the singularity of human and artificial intelligence (Eden et al., 2015). If well supervised, the collaboration is reciprocal and productive: artificial agents digitalize collaborative functioning, and human agents civilize their joint functioning (Yuste et al., 2017). In these respects, digitalization penetrates far deeper into human experience, compared to earlier phases of technological innovation. As Bandura (2006, p. 175) writes about the digital revolution, "These transformative changes are placing a premium on the exercise of agency to shape personal destinies and the national life of societies."

More specifically, digitalization is augmenting the sensory-perceptive, cognitive-affective, behavioral-performative, and evaluative-adaptive processes, which mediate human agency and personality (Mischel, 2004). In fact, artificial agents are being developed which imitate these features of human functioning. Enabling technologies will include artificial neural networks, quantum and cognitive computing, wearable computers, brain-machine engineering, intelligent sensors, and robotics. Smart digital assistants will also proliferate, reaching beyond smartphones to a wide range of digitally augmented interactions. These agents will deploy additional innovations, such as artificial personality and empathy (Kozma et al., 2018). Powered by such technologies, augmented agents will learn and act in far more expansive and effective ways. Consequently, a new type of agentic modality is emerging from digitalized human-machine collaboration.

Furthermore, assisted by digital technologies, people can more rapidly shift their attentional and calculative resources, updating memory, cognitive schema, and models of reasoning. In these respects, digital augmentation disrupts some traditional beliefs about the natural and human worlds. Particularly, given the massive growth of artificial intelligence and machine learning, the classic distinction between conscious mind and material nature appears unsustainable, as artificial agents become functionally sentient and empathic. Similarly, human collaboration with artificial personalities will challenge assumptions about privacy and the opacity of the self, because augmented agents will interpret and imitate empathy and other expressions of personality (Bandura, 2015). Therefore, a number of widely assumed distinctions appear increasingly contingent, and better viewed as options along a continuum, rather than as invariant categories (Klein et al., 2020). As Herbert Simon (1996), one of the founders of modern computer science and behavioral theory, observed, scientific insight often transforms assumed states into dynamic processes. In this case, insight transforms assumed material and conscious states into dynamic processes.

At the same time, there are grounds for concern and caution. To begin with, digitalization might enable new forms of oppression, superstition, and discrimination. Indeed, we already see evidence of these negative effects. For example, some institutional actors leverage digital technologies to dominate and oppress populations, for ideological, political, or commercial gain (Levy, 2018). Others use digital systems to restrict and distort information, spreading deliberate falsehood, superstition, and bias, again to serve self-interest. In addition, digitalization could be used to prolong the unsustainable, overexploitation of the natural world. Its benefits may also be unfairly distributed, privileging those who already possess capabilities and resources. Digitalization would then reinforce meritocratic privilege and undermine commitment to the common good (see Sandel, 2020). If this happens, the "digital divide" will continue to widen, exacerbating inequality across a range of social indicators, from mobility to education, health, political influence, and income. This book examines some of the underlying mechanisms which drive these effects.

Adaptive Challenges

Technological transitions of this scale are often fraught. They demand changes to fundamental beliefs and behaviors which are firmly encoded in culture and collective mind. Amending them is not easy. Nor should it be. Such beliefs and behaviors are typically contested and tested before encoding occurs, and the results are worthy of respect. Adding to the overall resilience of these systems, mental plasticity often declines with age, and most people adapt more slowly over time. Youthful curiosity and questioning give way to adult certainty and habit. Older institutions and organizations exhibit comparable tendencies. Although, here too, sluggish adaptation is sometimes advantageous. It may preserve evolutionary fitness in the face of temporary perturbation. In fact, without adequately stable ecologies, populations, and behaviors, biological and social order would neither evolve nor persist (Mayr, 2002). For this reason, incessant adaptation can be self-defeating or an early sign of impending ecological collapse.

Digital augmentation simultaneously compounds and disrupts this dynamic. Compounding occurs, because digital augmentation might lead to excessive adaptation and the unintended erosion of ecological stability and agentic fitness. In fact, without adequate supervision and constraint, psychosocial coherence could be at risk. At the same time, the sheer speed and power of these technologies can be disruptive. Many human systems are not designed for rapid change and might fracture under pressure. Furthermore, even if digitalization improves adaptive fitness, in doing so, it might shift the locus of control away from human agents, toward artificial sources. Hence, as artificial agents become more capable and ubiquitous, humanity must learn how to supervise its participation in augmented agency, while artificial agents must learn to incorporate human values, interests, and commitments, where commitment, in this context, means being dedicated, feeling obligated and bound to some value, belief, or pattern of action (Sen, 1985). Put simply, human agents need to digitalize, and artificial agents need to humanize. Many benefits are possible if digital augmentation enriches agentic capability and potentiality. If poorly supervised, however, artificial and human agents might diverge and conflict, even as they seek to collaborate. Or one agent may dominate the other and they will overly converge. Augmented humanity needs to understand and manage the resulting dilemmas.

New Problematics

In fact, digital augmentation problematizes modern assumptions about human capability and potentiality, where problematization is defined as raising new questions about the fundamental concepts, beliefs, and models of a field of enquiry (Alvesson & Sandberg, 2011). Thus defined, problematization looks beyond the refinement of existing theory. It is more than critique. It questions deeply held assumptions and invites the reformation of enquiry. For the same reason, problematization does not entail a detailed review of all prior work. Rather, we need to identify key concepts, assumptions, and models and then apply fresh thinking, all the while, reflecting on the novel phenomena and puzzles which prompt this process. My argument adopts such an approach. It problematizes modernity's core assumptions about human agentic capability and potentiality and examines the emerging problematics of digitally augmented humanity.

In short, modernity assumes that human agents are capable but limited, and need to overcome numerous constraints, to develop and flourish. As the preface to this work also states, modernity therefore focuses on the following questions: how can human agents collaborate with each other, in collective thought and action, while developing as autonomous persons; how can humans absorb change, while preserving value commitments and psychosocial coherence; how can societies develop stronger institutions and organizations, while avoiding the risks of excessive docility, determinism, and domination; how can humanity ensure fair access to the benefits of modernity and not allow growth to perpetuate discrimination, deprivation, and injustice; and finally, a defining challenge of modernity, asks to what degree, can and should human beings overcome their natural limits, to be more fully rational, empathic, and fulfilled (Giddens, 2013). As Kant (1964, p. 131) wrote regarding moral imperatives, we strive "to comprehend the limits of comprehensibility." Continuing this tradition, contemporary scholars investigate the limits of human understanding and how to transcend them, hoping to increase agentic capability and potential while balancing individual and collective priorities.

However, owing to digitalization, capabilities are expanding rapidly. Humans are potentially less limited, in many respects. Digitalization therefore leads us to problematize the modern assumption that human agency is inherently limited. New questions and problematics emerge instead. I also list these in the preface and repeat them here: how can human beings collaborate closely with artificial agents, while remaining genuinely autonomous in reasoning, belief, and choice; relatedly, how can humans integrate digital augmentation into their subjective and inter-subjective lives, while preserving personal identities, commitments, and psychosocial coherence; how can digitally augmented institutions and organizations, conceived as collective agents, fully exploit artificial capabilities, while avoiding extremes of digitalized docility, dependence, and determinism; how can humanity ensure fair access to the benefits of digital augmentation and not allow them to perpetuate systemic discrimination, deprivation, and injustice; and finally, the most novel and controversial challenge, which is how will human and artificial agents learn to understand, trust, and respect each other, despite their different levels of capability and potentiality. This is controversial because it implies that artificial agents will exhibit autonomous judgment and empathy. It assumes that sometime soon, we will attribute intentional agency to artificial agents (Ventura, 2019; Windridge, 2017).

1.1 Theories of Agency

Albert Bandura (2001) is a towering figure in the psychology of human agency, both individual and collective. His social cognitive theories explain how the capacity for self-regulated, self-efficacious action, is the hallmark of human agency, as well as a prerequisite for human self-generation and flourishing. In this respect, Bandura epitomizes the modern perspective on human agency: despite their natural limitations, people are capable of self-regulated, efficacious thought and action. They sense conditions in the world, identify and resolve problems, and pursue purposive goals. Human potential is thereby realized as people develop, engage in purposive action, and learn. They also mature as reflexive beings, acquiring the capability to monitor and manage their own thoughts and actions, and ultimately to self-generate a life course. Thus empowered and confident, people find fulfillment and flourish. In modernity, being truly human is to be freely and fully agentic.

Persons in Context

For comparable reasons, Bandura (2015) is among the psychologists who advocate situated models of human agency, personality, and rationality, often termed the "persons in context" and "ecological" perspectives. Like other scholars in this community, Bandura views human agency in naturalistic terms, assuming agents are sensitive to context, inherently variable, adaptive, and self-generative (Bandura, 2015; Bar, 2021; Cervone, 2004;

Fiedler, 2014; Kruglanski & Gigerenzer, 2011; Mischel & Shoda, 2010). Consequently, he and others reject static models of human personality and agency—for example, they reject fixed personality states and traits (e.g., McCrae & Costa, 1997)—and argue instead that persons are situated and adaptive. In the context of digitalization, conceiving of human agents in this way is important for two main reasons. First, if humans are complex, open, adaptive systems, situated, and self-generative, they are well suited for collaboration with artificial agents which share the same characteristics. Second, digitalization amplifies the impact of contextual dynamics because contexts change rapidly and penetrate more deeply into human experience. Being human in a digitalized world is to be human in augmented contexts.

For similar reasons, some psychologists explicitly compare human beings to artificial agents. They note that both types of agent can be modeled in terms of inputs, processes, and outputs (Shoda et al., 2002). In addition, both human and artificial agents sense the environment and gather information which they process using intelligent capabilities, leading to goal-directed action, and subsequent learning from performance. Humans and advanced artificial agents are both potentially self-generative as well. Therefore, human and artificial agents are deeply compatible, because both possess the same fundamental characteristics: (a) they are situated and responsive to context; (b) they use sensory perception of various kinds to sample the world and represent its problems; (c) both then apply intelligent processes to solve problems and develop action plans; (d) they self-regulate performances, including goal-directed action; (e) both evaluate performance processing and outputs, which results in learning, depending on sensitivity to variance; (f) both are self-generative and can direct their own becoming; and (g) they do all this as separate individual agents or within larger cooperative groups and networks.

Two of these characteristics are especially notable, namely generativity and contextuality. First, self-generation reflects a wider interest in generative processes broadly conceived. In numerous fields, scholars research how different systems kinds originate, produce, and procreate form and function without external direction. Chomsky's (1957) theory of generative grammar is a perfect example. In it, he argues that semantic principles are genetically encoded, then embodied in neurological structures, and subsequently generate linguistic systems. In this way, the first principles of grammar help to generate language. Others propose generative models of social science, using agent-based and neurocognitive modeling (e.g., Epstein, 2014). Some economists exploit the same methods to explain the origins and dynamics of markets (e.g., Chen, 2017; Dieci et al., 2018). While in personal life, generativity embraces the parenting of children, mentoring the young, as well as curating identities and life stories (McAdams et al., 1997).

Second, contextuality is not limited to theories of human personality and agency. For example, philosophers also debate the role of context, when considering the content of an agent's thoughts and actions. Not surprisingly, naturalistic and pragmatic philosophers are highly skeptical of ideal objectivity free from contextual influence. As Amartya Sen (1993) argues, perception, observation, belief, and value, all arise in some context and positions within it. From this perspective, claims of objectivity, whether ontological, epistemological, or ethical, must be positioned within context. There is no view from nowhere (see Nagel, 1989) and no godlike position or point of view, *sub specie aeternitatis*, which John Rawls (2001) hoped for. Commitments of every kind imply context and position. Human agents are forever situated, embedded in social, cultural, and historical contexts, although, each context and position can be well lit, by focused attention, sound reasoning, and gracious empathy.

In fact, across many fields of enquiry, scholars are adopting similar approaches. Context and position matter. Examples are found in other areas of psychology and social theory (Giddens, 1984; Gifford & Hayes, 1999), in economics (Sen, 2004), as well as in linguistics and discourse analysis (Lasersohn, 2012; Silk, 2016). They all share a common motivation. Within each field of enquiry, there is growing awareness of contextual variance and complexity, plus skepticism about static methods and models. These concerns are amplified by the obvious increase in phenomenal novelty and dynamism, especially owing to digitalization and related global forces. At the same time, most scholars who embrace context and position also reject unfettered subjectivity and relativism. Rather, they problematize assumptions about universals and ideals and investigate systematic processes of variation and adaptation instead. All agentic states are then conceived as processes in context unless there is compelling evidence to the contrary. Debate then shifts to what common, underlying systems or structures might exist among different expressions of agency. To cite Simon (1996) once again, scientific insight often transforms the understanding of assumed states into dynamic processes.

Capability and Potentiality

However, individual human agency is not simply an expression of personality in context. While agency assumes personality, it goes further (Bandura, 2006). The two constructs are not fully correlated. First, agency is forward looking, prospective, and aspirational, whereas personalities need not be. Second, agency is self-reactive, allowing agents to evaluate and respond to their own processes and performances. This function exploits outcome sensitivity and various feedback and feedforward mechanisms. Third, human agents are self-reflective, whereby they process information about their own states and performances and form reflexive beliefs and affects. Fourth, agency is potentially self-generative, meaning agents curate their own life path and way of becoming, although not all persons do so. To summarize, individual agency is an affordance of personality. The two are integrated, interdependent systems of human functioning. Personality and agency together, allow individuals to be intentional, prospective, aspirational, self-reactive, self-reflective, and self-generative.

Collective agents exhibit comparable characteristics. Yet collective agency is not simply the aggregation of personalities (Bandura, 2006). Granted, collectives connect and combine different individuals, but at the same time, collective agency is more holistic and qualitatively different. It relies heavily on networks and culture, for example, which also help to define collective modality and action (DiMaggio, 1997; Markus & Kitayama, 2003). Nevertheless, collectives share many of the same functional qualities as individuals. Collective agency is also intentional, prospective, aspirational, self-reactive, self-reflective, and self-generative. But these are now properties of communities, organizations, institutions, and networks, rather than individuals or aggregations of them (March & Simon, 1993; Scott & Davis, 2007). In summary, collective agency is an affordance of cultural community. In this respect, cultural communities and collective agency are integrated, interdependent systems of human functioning as well, but at a more complex level of organization and modality.

Limits of Capabilities

Irrespective of agentic modality and context, however, purely human capabilities are limited. Theories of agency therefore allow for approximate outcomes, trade-offs, and heuristics. They explain how individuals and collectives simplify and compromise, in order to reason and act within their limits (Gigerenzer, 2000; March & Simon, 1993). Sometimes, simplifying heuristics and trade-offs work well. But at other times, agents fall prey to noise, bias, and myopia, owing to the fallibility of such strategies (Fiedler & Wanke, 2009; Kahneman et al., 2016). Each major area of agentic processing is affected. First, sensory perception is constrained by limited attentional and observational capabilities, and agents easily misperceive the world and themselves, becoming myopic or clouded by noise. Second, cognitive-affective processes are limited by bounded calculative capabilities, which allow biases and myopias to distort problem-solving, decision-making, and preferential choice. Empathic capabilities are limited as well, meaning agents often struggle to interpret and understand other people and themselves. Third, behavioral-performative outputs are constrained by limited self-efficacy and self-regulatory capabilities. Hence, humans often perform poorly or inappropriately. And fourth, updates from feedforward and feedback are limited by insensitivity to variance, memory capacity, and procedural controls, meaning humans often fail to learn adequately and correctly. Feedforward updating is especially vulnerable, owing to its complexity and speed.

Importantly, these limitations suggest the contingency of many assumed criteria of reality, rationality, and justice (Bandura, 2006). For if purely human capabilities are inherently limited, then whatever is grounded in such capabilities will be limited as well. This is especially problematic, because ordinary categories and beliefs often acquire ideal status, as fundamental realities, necessary truths, and mandatory self-guides. They are idealized, meaning they are extrapolated to apply universally and forever, when in fact they do not (Appiah, 2017). Once again, each area of agentic processing is affected. First, the ordinary limits of sensory-perceptive capabilities often determine agents' fundamental ontological commitments and the core categories of reality. For this reason, most naturalistic and behavioral theorists argue that ontologies are contextual and variable to some degree, and hence open to revision (Gifford & Hayes, 1999; Quine, 1995). In contemporary philosophy, this approach supports "conceptual engineering," in which fundamental concepts of reality and value are constructed and reconstructed to fit the context (Burgess et al., 2020; Floridi, 2011).

Second, agents regularly hold idealized epistemological commitments—criteria of true belief and models of reasoning—which reflect the limits of their cognitive capabilities. Most naturalistic, and behavioral theories view epistemic commitments as inherently adaptive and ecological (Kruglanski & Gigerenzer, 2011). One notable advocate of this position was the later Wittgenstein (2009), who illuminated how contingent "language games" become idealized in axiomatic models of reasoning. In fact, Wittgenstein exposed axiomatic models as a type of meta-game, which foreshadowed recent thinking about the evolution of logics (e.g., Foss et al., 2012; Thornton et al., 2012). Third, agents adopt ethical commitments. They form ideals of goodness and justice which reflect their limited relational and empathic capabilities, where empathic limits constrain how much people can appreciate about each other's values and commitments. Philosophers then debate the origin of such limits and the degree to which they might be overcome. Some view empathic incompleteness as intractable and humanizing, and central to sociality and culture (e.g., Sen, 2009); while others argue for empathic universals, at least regarding fundamental principles (e.g., Rawls, 2001).

Impact of Digitalization

By transcending ordinary human capabilities, digital augmentation problematizes these questions and assumptions. First, digital innovations are rapidly improving the capacity to sense the environment, thereby heightening the perception of contextual variation and problems. Enabling technologies include the internet of things, intelligent sensing technologies, and fully autonomous agents. Second, digital augmentation massively increases information processing capabilities, transcending the assumed limits of human intelligence. For example, anyone with a contemporary smartphone can access enormous processing power at the touch of an icon. Third, digital augmentation enables new modes of action, which augment human performances. Digital innovations are transforming sophisticated domains of expert action, such as clinical medicine. Fourth, augmented agents can learn at unprecedented rates and degrees of precision, through rapid performance feedback, coupled with intense feedforward mechanisms (Pan et al., 2016; Pan & Yu, 2017).

Altogether, therefore, digitalization is radically augmenting agentic capabilities and potentialities, regarding sensory perception, cognitive-affective processing, behavior performance, evaluation of performance, and learning. In consequence, many traditional assumptions appear increasingly contingent and contextual, among them, conceptions of cognitive boundedness, distinctions between conscious mind and material nature, interpretive versus causal explanation, and abstract necessity versus practical contingency. Digitalization thus problematizes the conceptual architecture of modernity.

1.2 Metamodels of Agency

Fully to conceptualize and analyze this shift, we need to work at a higher level of metamodels. By way of definition, metamodels capture the common features of a related set of potential models within a field (Behe et al., 2014; Caro et al., 2014). Put simply, metamodels define families of models. They are specified by hyperparameters which define the core categories, relations, and mechanisms shared by a set of models (Feurer & Hutter, 2019). Thus defined, metamodels are studied in numerous fields, even if they are not labeled as such, for example, in decision-making (He et al., 2020; Puranam et al., 2015) and Chomsky's (2014) work on linguistics. The concept is very well established in computer science:

"Metamodels determine the set of valid models that can be defined with models' language and behavior in a particular domain" (Sangiovanni-Vincentelli et al., 2009, p. 55). In this book, the term refers to families, or related sets, of models of agency. Regarding agentic metamodels, hyperparameters define levels of organization or modality, activation mechanisms, and processing rates, such as the speed of self-regulation and learning, where rates, in this context, are defined as the number of processing cycles performed per unit of time. The reader will therefore encounter the terms "metamodel of agency" and "agentic metamodel" throughout this book. However, I will not offer an alternative model of agency at the detailed level. The book does not present an alternative theory of human psychology or agency as such. Nor will it propose a formal model of augmented agency or humanity based on a specific theory. Rather, my argument will focus at a higher level, on the features of agentic metamodels.

To illustrate, consider the field of psychological science. In this field, a popular metamodel assumes that human beings perceive the world and themselves, then process information and perform action, with varying degrees of intelligence and autonomy. Human beings are therefore a type of input-process-output system (Mischel, 2004). Given this broad metamodel, scholars then formulate specific models of psychological functioning, such as behaviorist, social cognitive, state, and trait models. Importantly, each type of model exemplifies the principles of the broad metamodel, though they vary in terms of the specific parameters for inputs, the internal mechanisms of processing, and performance outputs. Moreover, in fields like psychology, domain-specific metamodels are often predetermined, typically from the analysis of practice and experience. Indeed, whole industries evolve this way. Prescriptive metamodels guide pedagogical and clinical practice (Bandura, 2017). Most psychologists therefore assume fairly stable agentic metamodels which are deeply encoded in culture and community (Soria-Alcaraz et al., 2017). This means that metamodels adapt incrementally, if at all, under normal conditions. Indeed, institutional fields are labeled "fields" for this reason; and similarly, personality types are labeled "types." Both labels reflect stable metamodels in these fields of study (Mischel, 2004; Scott, 2014). Moreover, few practitioners question the normative metamodels of a field. Most are encoded during training, or imposed by regulation, and remain fixed.

The question remains, however, whether metamodels will help in the analysis of human-artificial augmented agency. Even if metamodeling is a suitable way to analyze both human and artificial agents, at a high level, can the two be integrated in this way? Perhaps the fundamental features of the mind and consciousness are too incommensurable with artificial agency and intelligence. Arguably, this was the case until recently. However, as noted earlier, recent technical advances suggest that metamodeling is now feasible in this regard. For example, advanced systems of artificial intelligence are increasingly capable of higher forms of cognitive functioning, including self-generation and self-supervision, associative and speculative reasoning, heuristic problem-solving and decision-making, as well as interpreting affect and empathy (Asada, 2015; Caro et al., 2014). Human and artificial agents are increasingly similar and thus amenable to integrative metamodeling, especially when they combine as augmented agents.

Compositive Methods

Digitalized ecologies will be increasingly dynamic and responsive. Agency will be less reliant on stable metamodels and encoded templates. New metamodels, or families of models, will consistently emerge. In this way, augmented agents will be capable of rapid transformation. Humans and artificial agents will take on different, complementary roles, self-generating dynamically to fit changing contexts. Their metamodels will compose and recompose in real time, to fit changing conditions. In this respect, digital augmentation supports a more dynamic method, which can be described as "compositive" (cf. Latour, 2010), meaning that methods and models will compose, decompose, or recompose, to fit different contexts. From a design perspective, therefore, augmented agency will be near composability, as well as being near decomposable modular and hierarchical systems. Moreover, compositive methods are systematic and rigorous, the result of processing vast quantities of data. These methods are neither ad hoc nor idiosyncratic (e.g., Pappa et al., 2014; Wang et al., 2015).

Compositive methods are already employed in contemporary artificial intelligence. Systems maintain databases of processing modules and methods, and then select and combine these to fit the problem context. Metamodels and models are developed rapidly, contextually, in response to problems and situations. As noted previously, the most advanced software algorithms now compose their own metamodels— they are fully self-generative—requiring minimal (if any) supervision. Evolutionary deep learning systems and Generative Adversarial Networks (GANs) function in exactly this way (Shwartz-Ziv & Tishby, 2017). Via rapid inductive and abductive learning, these systems process massive volumes of information, identifying hitherto undetectable patterns, to compose new metamodels and models, often without any external supervision. Augmented agents will do likewise. They will leverage the power of digitalization to select and combine different techniques and procedures, and thereby compose metamodels and methods which best fit the context. Development of, and investigation by, augmented agents will invoke compositive methods.

Notably, the great economist, Friedrich Hayek (1952), argued for compositive methods in the social sciences, as an antidote to naïve reductionism, developing models and methods which best fit the problem at hand (Lewis, 2017). In these respects, Hayek's conception of "compositive" is comparable to recent technical developments. Going beyond Hayek's conception, however, digitalized metamodeling is agentic and ecological, more similar to Latour's (2010) concept of composition. It synthesizes both top-down and bottom-up processing, detailed and holistic, rapidly iterating, using prospective metamodeling and testing, until maximizing metamodel fit, and often achieving this in a fully unsupervised, self-generative fashion. In these respects, digitalized composition also problematizes traditional methodological distinctions: between qualitative and quantitative, methodological individualism and collectivism, and between reductionism and holism. Instead, compositive methods will blend these options and treat such polarities as the extremities of continua (Latour, 2011). I will return to these topics in later chapters, and especially in the final chapter which discusses the future science of digitally augmented agency.

1.3   Dimensions of Metamodeling

Nevertheless, given the complexity of many problems and the processing they require, artificial agents must also simplify and approximate. This is done using algorithmic heuristics, which are shortcut means of specifying models and methods (Boussaid et al., 2013). At the most general level, hyperheuristics provide simplified means of specifying the broad hyperparameters of metamodels. Recall that metamodels are defined as related sets of potential models, and hyperparameters specify the broad features or attributes of metamodels, including their core categories, mechanisms, and processing rates (Feurer & Hutter, 2019). Hyperheuristics are shortcut means of specifying these properties. Metamodels are further distinguished by the supervision applied in their development. As noted earlier, they can be fully supervised, semi-supervised, or unsupervised, from artificial and human sources.

In supervised metamodeling, hyperparameters are fully determined by prior experience and learning (e.g., Amir-Ahmadi et al., 2018), whereas in semi-supervised systems, the initial hyperparameters are partially given, but provisional. Additional processing is required to tune and optimize them. Among the benefits of a semi-supervised approach, is that metamodeling can exploit prior learning while responding to novelty, although, semi-supervised metamodeling also poses risks, if it imports distorting biases and myopias (Horzyk, 2016). Alternatively, some artificial agents are fully unsupervised. Hyperparameters are developed by the agent itself, in a self-generative fashion. Metamodels are composed, rather than retrieved. Advanced artificial agents do this through rapid, iterative hyperparameter pruning, tuning, and optimization (Song et al., 2019).

As noted above, GANs are a recent innovation of this kind (Wang et al., 2017). In these systems, artificial agents compete in a collaborative game. A generator produces fake examples of some phenomenon, derived from pure noise. In parallel, a discriminator is trained on real examples of phenomena, such as photographs of human faces. If the system is fully unsupervised, these training data are unlabeled and unstructured. Using such data, the discriminator learns via multiple cycles of induction. Then the artificially generated, fake examples are passed to the discriminator, along with unclassified real examples, and the discriminator tries to distinguish real from fake. The competition ends in a Nash equilibrium, being the state in which neither the generator nor discriminator can do better against the other, but they rely on each other to achieve this maximal state, and both therefore benefit from stabilizing the system (Pan et al., 2019). In this fashion, the GAN produces a maximizing solution to the focal problem, for example, developing an artificial agent which can distinguish human faces, without needing any external supervision (Liong et al., 2020). And the metamodel is fully unsupervised and self-generative.

Parameters and Variables

Given initial hyperparameters and the metamodel they define, the next phase applies metaheuristics to select a model from the choice set (Feurer & Hutter, 2019). First, the agent will select specific parameters, about what counts as real versus fake, and what is exposed or hidden. Second, it will select activation functions, such as the type of action generation, or the outcome variance which triggers adaptive feedback. Third, there will be specific processing cycles and learning rates, for example, whether a particular type of feedback is slow and sluggish, or fast and hyperactive, and also about the level and intensity of feedforward processing. In purely human processes, such parameters tend to be encoded in memory and mental models, and supervised by metacognition (Bandura, 2017). Even ecological models of rationality are significantly supervised, by prescribing criteria of adaptation and association (Kruglanski & Gigerenzer, 2011). For example, the metaheuristic may encode "fast and frugal heuristics" as the most ecologically appropriate model for problem-solving (Gigerenzer & Goldstein, 1996). The system then employs this specific model to resolve a focal problem.

Next, given a specific model and its parameters, the process specifies the variables or expected patterns of variance. For example, in a naturalistic model of agency, variables might capture the expected degree and rate of variation in self-regulated behavior (the dependent variable), conditional on the strength or weakness of self-efficacy (the independent variable) (Bandura, 1997). In this case, the variables are supervised and predetermined. Though, if supervision is poor, the specification of variables can easily import distorting myopias and biases. Indeed, such biases often infect human and machine learning, resulting in poor choices and decision-making (Noble, 2018). Human and artificial agents must therefore learn how better to supervise the selection of variables, mindful of these risks. Otherwise, models could be underfitting (admitting too much noise and variance), or overfitting (excluding too much noise and variance). Both scenarios will increase functional losses (Kahneman et al., 2016).

Importantly, at each level of artificial processing, algorithmic heuristics help to manage the otherwise overwhelming complexity of data and processes. Indeed, much research into artificial intelligence and machine learning focuses on optimizing such hierarchies: using hyperheuristics to select the hyperparameters which define metamodel choice sets; then using metaheuristics to select the detailed model which fits best; and finally, the chosen model provides specific heuristics to solve a focal problem. The earlier example cited (a) the metamodel of associative, heuristic problem-solving, then (b) the model of "fast and frugal" heuristics, and (c) applying a specific heuristic, such as a simple stopping rule (Gigerenzer, 2000). These methods will be critical for the effectiveness and efficiency of digitalized problem-solving, and especially more complex problems. For the same reasons, these methods will be employed by digitally augmented agents.

Furthermore, depending on the type and level of supervision, hyperparameters are more, or less visible. Recall that some are predetermined, given by supervision, and hence immediately visible. Others may be hidden and unsupervised, and therefore wait to be discovered by further processing. In computer science, these questions loom large for the efficiency of artificial agents (Yao et al., 2017). On the one hand, the more prior supervision of hyperparameters, the less is hidden and metamodeling is more efficient and predictable. For example, in fully supervised machine learning, hyperparameters are predetermined and thus fully visible. However, as a result, there are fewer degrees of freedom: the greater the supervision, the less freedom in metamodeling. On the other hand, with less or no supervision, more is hidden. This entails greater degrees of freedom, to explore and self-generate. This is the case in unsupervised GANs, in which hyperparametric values are largely hidden and await discovery. However, the process of discovery consumes time and resources. To compensate, unsupervised systems also employ hyperheuristics in hyperparameter tuning and pruning, to optimize metamodel discovery and design. That is, they self-supervise their own objective function to maximize fit while also minimizing the processing load (Burke et al., 2019). Similar dynamics occur in the development and functioning of agentic metamodels. Human systems also need to balance metamodel fit and efficiency. But in these contexts, components can be hidden for other reasons, and especially owing to the limitations of human perception and consciousness.

The Role of Consciousness

In premodern cultures, it was assumed that most fundamental principles are accessible to ordinary consciousness, even if they depended on divine revelation and ritual. This included the core categories of reality, truth, and value, about persons, the polis, and the cosmos (Rochat, 2009). However, hyperparameters of this kind are inevitably anthropomorphic, owing to their origins in ordinary experience. This was certainly true for premodernity. Fundamental categories of reality and truth were defined in human terms, that is, in terms which reflected ordinary consciousness. Hence, the gods were superhuman characters and the cosmos emerged through anthropomorphic or animistic stories of creation. By implication, premodern cultures offered few degrees of self-generative freedom in agentic form and function.

By contrast, during post-Enlightenment modernity, the fundamental properties of nature are largely inaccessible to ordinary consciousness. To discover them, one requires specialized technological assistance, or in other words, the methods of modern empirical science. Nevertheless, many continued to believe that the fundamental properties of mind and self are directly accessible to consciousness. Descartes (1998) exemplified this belief when he introspected and famously concluded, "I think therefore I am." The modern mind-body problem was born and over time modernity bifurcated the sciences. On the one hand, the natural sciences demoted ordinary consciousness and relied on technological assistance to access the hidden, fundamental realities of nature. Whereas on the other hand, many human sciences continued relying on ordinary consciousness to access the fundamental properties of mind and self, with or without technological assistance (Thiel, 2011). Some disciplines continue to do so, believing that the hyperparameters of cognitive form and function are directly accessible to ordinary consciousness. Arguably, this is anthropomorphic and erroneous (see Chomsky, 2014).

In fact, owing to digitalization and neurophysiological discoveries, it is becoming abundantly clear that the fundamental realities of mind and self are opaque to ordinary consciousness (Carruthers, 2011). Specialized technologies are required here too. Introspection is a functional approximation, at best. From the perspective of digital augmentation, therefore, no fundamental categories and mechanisms—of neither physical nature nor mental phenomena—are directly accessible to ordinary consciousness. Both require technological assistance to observe and analyze them. However, this does not entail the reduction of mind and self to material cause or the digital dissolution of consciousness. Rather, as I will explain more fully in later chapters, it entails rethinking classic concepts of mind and self in terms of digitally augmented agency and self-generative systems.

Significant implications follow for the supervision of technologically assisted agency, and especially the supervision of digitally augmented agents. Most importantly, if ordinary consciousness is demoted and no longer a reliable source of fundamental reality and truth, then it will require deliberate supervision to ensure that ordinary human inputs are acknowledged and respected. They cannot, and should not, be either foundational or taken for granted. In fact, this problem is already a topic of research in computer science. Artificial agents are designed to recognize and accommodate the ordinary experience of mind and self, when they need to collaborate with humans in behavioral settings (Abbass, 2019), for example, when humans travel in autonomous vehicles. These situations require the systematic incorporation of human perceptions, values, and interests, despite their lack of precision and reliability. In this way, the supervision of augmented agency is humanized.

The earlier Figs. 1.1 and 1.2 illustrate these effects. Recall that these figures depict the core supervisory challenge of technologically assisted humanity, namely, how to combine and coordinate, divergent levels of human and technological functionality. In fact, the same factors explain the shifting role of ordinary consciousness in explanatory thought. To illustrate, instead of interpreting these figures as general models of supervision, now assume they depict the supervision of explanatory thought. Next, recall that the small gap between levels of capability L_1 and L_2 in Fig. 1.1 illustrates modest technological assistance. We can therefore reinterpret this figure to depict forms of science with modest technological tools and techniques. Also, note that segments A and B are both relatively small. Much supervision is achievable using baseline capability L_1 and hence accessible to consciousness. In fact, this was the dominant pattern in premodern science (Sorabji, 2006). It persists in some fields of human study, which still derive fundamental categories and mechanisms from ordinary consciousness and introspection.

By contrast, in Fig. 1.2, there is a larger gap between baseline capability L_1 and more technologically advanced capabilities L_3. This figure therefore illustrates forms of science with significant technological input. Segments C and D are large, implying that much is inaccessible to ordinary consciousness and requires supervision at level L_3. Modern natural science is certainly like this, as are the human sciences which no longer rely on ordinary consciousness and perception but employ specialized technologies instead. The science of digitally augmented agency will adopt the same approach. However, for this reason, future science confronts a major challenge. It will require strong collaborative supervision to avoid scenarios in which artificial agents overwhelm and ignore human inputs, and/or human supervision imports distorting myopias and bias into artificial intelligence and science.

Critical Dilemmas

Metamodeling therefore plays an important role in agentic thought and action. In practical, behavioral domains, most metamodeling is automatic, implicit, encoded in memory, and heavily supervised by procedural routine and custom. Data are labeled and principles are clear. In fact, in ordinary human experience, metamodeling is only self-generative in the most creative and speculative domains. By contrast, metamodeling by artificial agents is increasingly self-generative and unsupervised. This distinction between human and artificial supervision of metamodeling has profound implications for their collaboration. Digitally augmented humanity must integrate both types of agent and accommodate their different capabilities and potentialities. On the one hand, human inputs will be strongly supervised, replicative, and ordinary human intuitions and priors will often persist. Humans also tend to be comparatively myopic, sluggish, layered, and insensitive to variance. Whereas artificial agents will tend toward increasingly unsupervised, self-generated inputs, independent of human intervention. In addition, artificial agents are comparatively farsighted, fast, compressed, and hypersensitive to variance.

Overall, collaborative supervision will therefore be daunting, even for the most developed augmented agents (Cheng et al., 2020). If supervision is poor, the result could be extremely convergent or divergent forms and functions. Regarding over-convergence, one type of agent might dominate the other resulting in systems which are too digitalized or too humanized. Whereas regarding over-divergence, human and artificial inputs will both be significant but conflicting. A number of divergent dilemmas are possible. First, the human and artificial components of augmented agents could diverge in terms of range, being both farsighted and nearsighted at the same time, looking too far and near in sampling and search. Second, their processing rates might diverge, being rapid in some respects and sluggish in others, thus cycling both too fast and too slow. Third, artificial processes could be hypersensitive to variance, while human processes are relatively insensitive, thereby admitting too much and too little noise. And fourth, augmented agents might combine overly complex and simplified components, leading to poor integration and coordination. In all these scenarios, outcomes will easily become dysfunctional. The following chapters examine the origins and consequences of these dilemmas for key domains of agentic form and functioning. The final chapter looks forward to the future science of digitally augmented agency.

References

Abbass, H. A. (2019). Social integration of artificial intelligence: Functions, automation allocation logic and human-autonomy trust. *Cognitive Computation, 11*(2), 159–171.

Alvesson, M., & Sandberg, J. (2011). Generating research questions through problematization. *Academy of Management Review, 36*(2), 247–271.

Amir-Ahmadi, P., Matthes, C., & Wang, M.-C. (2018). Choosing prior hyper-parameters: With applications to time-varying parameter models. *Journal of Business & Economic Statistics, 38*(1), 124–136.

Appiah, K. A. (2017). *As if: Idealization and ideals.* Harvard University Press.

Asada, M. (2015). Towards artificial empathy. *International Journal of Social Robotics, 7*(1), 19–33.

Bandura, A. (1997). *Self-efficacy: The exercise of control.* W.H.Freeman and Company.

Bandura, A. (2001). Social cognitive theory: An agentic perspective. *Annual Review of Psychology, 52*, 1–26.

Bandura, A. (2006). Toward a psychology of human agency. *Perspectives on Psychological Science, 1*(2), 164–180.

Bandura, A. (2007). Reflections on an agentic theory of human behavior. *Tidsskrift-Norsk Psykologforening, 44*(8), 995.

Bandura, A. (2015). On deconstructing commentaries regarding alternative theories of self-regulation. *Journal of Management, 41*(4), 1025–1044.

Bandura, A. (Ed.). (2017). *Psychological modeling: Conflicting theories.* Transaction Publishers.

Bar, M. (2021). From objects to unified minds. *Current Directions in Psychological Science, 30*(2), 129–137.

Behe, F., Galland, S., Gaud, N., Nicolle, C., & Koukam, A. (2014). An ontology-based metamodel for multiagent-based simulations. *Simulation Modelling Practice and Theory, 40*, 64–85.

Boussaid, I., Lepagnot, J., & Siarry, P. (2013). A survey on optimization meta-heuristics. *Information Sciences, 237*(Supplement C), 82–117.

Braudel, F., & Mayne, R. (1995). *A history of civilizations.* Penguin Books.

Burgess, A., Cappelen, H., & Plunkett, D. (Eds.). (2020). *Conceptual engineering and conceptual ethics.* Oxford University Press.

Burke, E. K., Hyde, M. R., Kendall, G., Ochoa, G., Ozcan, E., & Woodward, J. R. (2019). A classification of hyper-heuristic approaches: Revisited. In *Handbook of metaheuristics* (pp. 453–477). Springer.

Caro, M. F., Josyula, D. P., Cox, M. T., & Jimenez, J. A. (2014). Design and validation of a metamodel for metacognition support in artificial intelligent systems. *Biologically Inspired Cognitive Architectures, 9*, 82–104.

Carruthers, P. (2011). *The opacity of mind: An integrative theory of self-knowledge.* Oxford University Press.

Castaldi, C., & Dosi, G. (2006). The grip of history and the scope for novelty: Some results and open questions on path dependence in economic processes. In *Understanding change* (pp. 99–128). Springer.

Cervone, D. (2004). The architecture of personality. *Psychological Review, 111*(1), 183–204.

Chen, S.-H. (2017). *Agent-based computational economics: How the idea originated and where it is going.* Routledge.

Cheng, Q., Li, H., Wu, Q., & Ngan, K. N. (2020). Hybrid-loss supervision for deep neural network. *Neurocomputing, 388*, 78–89.

Chomsky, N. (1957). *Syntactic structures.* Mouton & Co.

Chomsky, N. (2014). *The minimalist program.* MIT Press.

Crone, P. (2015). *Pre-industrial societies: Anatomy of the pre-modern world.* Simon and Schuster.

Descartes, R. (1998). *Meditations and other metaphysical writings* (D. M. Clarke, Trans.). Penguin.

Dieci, R., He, X.-Z., Hommes, C., & LeBaron, B. (2018). Chapter 5 – Heterogeneous agent models in finance. In *Handbook of computational economics* (Vol. 4, pp. 257–328). Elsevier.

DiMaggio, P. (1997). Culture and cognition. *Annual Review of Sociology, 23*, 263–287.

Eden, A. H., Moor, J. H., Søraker, J. H., & Steinhart, E. (2015). *Singularity hypotheses.* Springer.

Epstein, J. M. (2014). *Agent_zero: Toward neurocognitive foundations for generative social science* (Vol. 25). Princeton University Press.

Feurer, M., & Hutter, F. (2019). Hyperparameter optimization. In *Automated machine learning* (pp. 3–33). Springer.

Fiedler, K. (2014). From intrapsychic to ecological theories in social psychology: Outlines of a functional theory approach. *European Journal of Social Psychology, 44*(7), 657–670.

Fiedler, K., & Wanke, M. (2009). The cognitive-ecological approach to rationality in social psychology. *Social Cognition, 27*(5), 699–732.

Floridi, L. (2011). A defence of constructionism: Philosophy as conceptual engineering. *Metaphilosophy, 42*(3), 282–304.

Foss, N. J., Heimeriks, K. H., Winter, S. G., & Zollo, M. (2012). A Hegelian dialogue on the micro-foundations of organizational routines and capabilities. *European Management Review, 9*(4), 173–197.

Geertz, C. (2001). *Available light*. Princeton University Press.

Giddens, A. (1984). *The constitution of society*. University of California Press.

Giddens, A. (2013). *The consequences of modernity*. Wiley.

Gifford, E. V., & Hayes, S. C. (1999). Functional contextualism: A pragmatic philosophy for behavioral science. In *Handbook of behaviorism* (pp. 285–327). Elsevier.

Gigerenzer, G. (2000). *Adaptive thinking: Rationality in the real world*. Oxford University Press.

Gigerenzer, G., & Goldstein, D. G. (1996). Reasoning the fast and frugal way: Models of bounded rationality. *Psychological Review, 103*(4), 650–669.

Hayek, F. A. (1952). *The counter-revolution of science: Studies on the abuse of reason*. The Free Press.

He, L., Zhao, W. J., & Bhatia, S. (2020). An ontology of decision models. *Psychological Review* (online).

Horzyk, A. (2016). Human-like knowledge engineering, generalization, and creativity in artificial neural associative systems. In *Knowledge, information and creativity support systems: Recent trends, advances and solutions* (pp. 39–51). Springer.

Johnson-Laird, P. N. (2010). Mental models and human reasoning. *Proceedings of the National Academy of Sciences, 107*(43), 18243–18250.

Kahneman, D., Rosenfield, A. M., Gandhi, L., & Blaser, T. (2016). Noise: How to overcome the high, hidden cost of inconsistent decision making. *Harvard Business Review, 94*(10), 38–46.

Kant, I. (1964). *Groundwork of the metaphysic of morals* (H. J. Paton, Trans.). Harper & Row.

Klein, H. J., Solinger, O. N., & Duflot, V. (2020). Commitment system theory: The evolving structure of commitments to multiple targets. *Academy of Management Review* (online).

Kozma, R., Alippi, C., Choe, Y., & Morabito, F. C. (Eds.). (2018). *Artificial intelligence in the age of neural networks and brain computing*. Academic Press.

Kruglanski, A., & Gigerenzer, G. (2011). Intuitive and deliberate judgments are based on common principles. *Psychological Review, 118*(1), 97–109.

Lasersohn, P. (2012). Contextualism and compositionality. *Linguistics and Philosophy, 35*(2), 171–189.

Latour, B. (2010). An attempt at a "compositionist manifesto". *New Literary History, 41*(3), 471–490.

Latour, B. (2011). From multiculturalism to multinaturalism: What rules of method for the new socio-scientific experiments? *Nature and Culture, 6*(1), 1–17.

Latour, B. (2017). Why Gaia is not a god of totality. *Theory, Culture & Society, 34*(2–3), 61–81.

Lenski, G. (2015). *Ecological-evolutionary theory: Principles and applications.* Routledge.

Leonardi, P. M., & Barley, S. R. (2010). What's under construction here? Social action, materiality, and power in constructivist studies of technology and organizing. *Academy of Management Annals, 4*(1), 1–51.

Levy, F. (2018). Computers and populism: Artificial intelligence, jobs, and politics in the near term. *Oxford Review of Economic Policy, 34*(3), 393–417.

Lewis, P. A. (2017). Ontology and the history of economic thought: The case of anti-reductionism in the work of Friedrich Hayek. *Cambridge Journal of Economics, 41*(5), 1343–1365.

Liong, S.-T., Gan, Y. S., Zheng, D., Li, S.-M., Xu, H.-X., Zhang, H.-Z., Lyu, R.-K., & Liu, K.-H. (2020). Evaluation of the spatio-temporal features and GAN for micro-expression recognition system. *Journal of Signal Processing Systems, 92,* 705–725.

March, J. G., & Simon, H. (1993). *Organizations* (2nd ed.). Blackwell.

Markus, H. R., & Kitayama, S. (2003). Culture, self, and the reality of the social. *Psychological Inquiry, 14*(3/4), 277–283.

Markus, H. R., & Kitayama, S. (2010). Cultures and selves: A cycle of mutual constitution. *Perspectives on Psychological Science, 5*(4), 420–430.

Mayr, E. (2002). *What evolution is.* Weidenfeld & Nicolson.

McAdams, D. P., Diamond, A., de St. Aubin, E., & Mansfield, E. (1997). Stories of commitment: The psychosocial construction of generative lives. *Journal of Personality & Social Psychology, 72*(3), 678–694.

McCrae, R. R., & Costa, P. T., Jr. (1997). Personality trait structure as a human universal. *American Psychologist, 52*(5), 509–516.

Mischel, W. (2004). Toward an integrative science of the person. *Annual Review of Psychology, 55,* 1–22.

Mischel, W., & Shoda, Y. (2010). The situated person. In B. Mesquita, L. F. Barrett, & E. R. Smith (Eds.), *The mind in context* (pp. 149–173). Guilford Press.

Murray, A., Rhymer, J., & Sirmon, D. G. (2020). Humans and technology: Forms of conjoined agency in organizations. *Academy of Management Review* (online).

Nagel, T. (1989). *The view from nowhere*. Oxford University Press.

Noble, S. U. (2018). *Algorithms of oppression: How search engines reinforce racism*. NYU Press.

Norvig, P., & Russell, S. (2010). *Artificial intelligence: A modern approach* (3rd ed.). Pearson.

Pan, Y., & Yu, H. (2017). Biomimetic hybrid feedback feedforward neural-network learning control. *IEEE Transactions on Neural Networks and Learning Systems, 28*(6), 1481–1487.

Pan, Y., Liu, Y., Xu, B., & Yu, H. (2016). Hybrid feedback feedforward: An efficient design of adaptive neural network control. *Neural Networks, 76*, 122–134.

Pan, Z., Yu, W., Yi, X., Khan, A., Yuan, F., & Zheng, Y. (2019). Recent progress on generative adversarial networks (GANs): A survey. *IEEE Access, 7*, 36322–36333.

Pappa, G. L., Ochoa, G., Hyde, M. R., Freitas, A. A., Woodward, J., & Swan, J. (2014). Contrasting meta-learning and hyper-heuristic research: The role of evolutionary algorithms. *Genetic Programming and Evolvable Machines, 15*(1), 3–35.

Pinker, S. (2018). *Enlightenment now: The case for reason, science, humanism, and progress*. Penguin.

Puranam, P., Stieglitz, N., Osman, M., & Pillutla, M. M. (2015). Modelling bounded rationality in organizations: Progress and prospects. *The Academy of Management Annals, 9*(1), 337–392.

Quine, W. V. (1995). Naturalism; or, living within one's means. *Dialectica, 49*(2–4), 251–263.

Rawls, J. (2001). *A theory of justice* (Revised ed.). Harvard University Press.

Rochat, P. (2009). *Others in mind: Social origins of self-consciousness*. Cambridge University Press.

Sandel, M. J. (2020). *The tyranny of merit: What's become of the common good?* Penguin Books.

Sangiovanni-Vincentelli, A., Shukla, S. K., Sztipanovits, J., Yang, G., & Mathaikutty, D. A. (2009). Metamodeling: An emerging representation paradigm for system-level design. *IEEE Design & Test of Computers, 26*(3), 54–69.

Scott, W. R. (2014). *Institutions and organizations: Ideas, interests, and identities* (4th ed.). Sage.

Scott, W. R., & Davis, G. F. (2007). *Organizations and organizing: Rational, natural and open system perspectives*. Pearson Education.

Sen, A. (1985). Goals, commitment, and identity. *Journal of Law, Economics & Organization, 1*(2), 341–355.

Sen, A. (1993). Positional objectivity. *Philosophy & Public Affairs, 22*(2), 126–145.

Sen, A. (2004). Economic methodology: Heterogeneity and relevance. *Social Research, 71*(3), 583–614.

Sen, A. (2009). *The idea of justice.* Harvard University Press.

Shoda, Y., LeeTiernan, S., & Mischel, W. (2002). Personality as a dynamical system: Emergence of stability and distinctiveness from intra- and interpersonal interactions. *Personality and Social Psychology Review, 6*(4), 316–325.

Shwartz-Ziv, R., & Tishby, N. (2017). Opening the black box of deep neural networks via information. *arXiv preprint arXiv:1703.00810.*

Silk, A. (2016). *Discourse contextualism: A framework for contextualist semantics and pragmatics.* Oxford University Press.

Simon, H. A. (1979). *Models of thought* (Vol. 352). Yale University Press.

Simon, H. A. (1996). *The sciences of the artificial* (3rd ed.). The MIT Press.

Smith, A. (1950). *An inquiry into the nature and causes of the wealth of nations (1776).* Methuen.

Smith, R. L. (2008). *Premodern trade in world history.* Routledge.

Song, H., Triguero, I., & Ozcan, E. (2019). A review on the self and dual interactions between machine learning and optimisation. *Progress in Artificial Intelligence, 8*(2), 143–165.

Sorabji, R. (2006). *Self: Ancient and modern insights about individuality, life, and death.* OUP.

Soria-Alcaraz, J. A., Ochoa, G., Sotelo-Figeroa, M. A., & Burke, E. K. (2017). A methodology for determining an effective subset of heuristics in selection hyper-heuristics. *European Journal of Operational Research, 260*(3), 972–983.

Spar, D. L. (2020). *Work mate marry love: How machines shape our human destiny.* Farrar, Straus and Giroux.

Thiel, U. (2011). *The early modern subject: Self-consciousness and personal identity from Descartes to Hume.* Oxford University Press.

Thornton, P. H., Ocasio, W., & Lounsbury, M. (2012). *The institutional logics perspective: A new approach to culture, structure, and process.* Oxford University Press.

Ventura, D. (2019). Autonomous intentionality in computationally creative systems. In *Computational creativity* (pp. 49–69). Springer.

Wang, B., Xu, S., Yu, X., & Li, P. (2015). Time series forecasting based on cloud process neural network. *International Journal of Computational Intelligence Systems, 8*(5), 992–1003.

Wang, K., Gou, C., Duan, Y., Lin, Y., Zheng, X., & Wang, F. (2017). Generative adversarial networks: Introduction and outlook. *IEEE/CAA Journal of Automatica Sinica, 4*(4), 588–598.

Wilson, E. O. (2012). *The social conquest of earth*. WW Norton & Company.

Windridge, D. (2017). Emergent intentionality in perception-action subsumption hierarchies. *Frontiers in Robotics and AI, 4*, 38.

Wittgenstein, L. (2009). *Philosophical investigations*. Wiley.

Yao, C., Cai, D., Bu, J., & Chen, G. (2017). Pre-training the deep generative models with adaptive hyperparameter optimization. *Neurocomputing, 247*, 144–155.

Yuste, R., Goering, S., Bi, G., Carmena, J. M., Carter, A., Fins, J. J., Friesen, P., Gallant, J., Huggins, J. E., & Illes, J. (2017). Four ethical priorities for neurotechnologies and AI. *Nature News, 551*(7679), 159.

2

Historical Metamodels of Agency

Different agentic metamodels correspond to major historical periods of civilized humanity. Technological innovation is an important, distinguishing feature of this narrative (Spar, 2020). The long-term trend is toward greater agentic capability and potentiality, assisted by more sophisticated technologies: from the static agentic metamodels and simple technologies of premodernity to the more complex metamodels of modernity, assisted by mechanical and analogue technologies, and now to the increasingly dynamic, digitally augmented metamodels of the contemporary period. In summary, the evolution of agentic metamodels is a historical process itself. Each major period warrants detailed discussion.

2.1 Major Historical Periods

In premodern cultures—prior to the modern period of Enlightenment and industrialization—human agency was popularly conceived in terms of divinely ordained narratives and fixed social orders. Purposive thought and action were supervised by patriarchal authority and supernatural

P. T. Bryant, *Augmented Humanity*, https://doi.org/10.1007/978-3-030-76445-6_2

beings, which helped to make sense of intractable fate. For most people, the order of things was not a human composition, but bestowed by agents from above and beyond (Geertz, 2001). Given these assumptions, explanation of the world was teleological and driven by final cause, while categories of reality were defined in terms of essential states and forms. Reflecting this relative lack of capability and potentiality, the dominant metamodels of agency assumed stability (Sorabji, 2006). Normality meant replicating an established, often divinely ordained metamodel of agency, and significant variance was viewed as a sign of weakness or deviance. Hence, trying to amend or circumvent the divine order was fraught with existential risk, as the ancient Greek tragedians understood (Williams, 1993). And for most, human overcoming was not explained by autonomous reasoning and action, but by good fortune and supernatural beneficence.

In like fashion, explanatory thought about agency in premodernity focused on collective norms, compliance with them, and the vicissitudes of fate. Rare opportunities for change consisted in altering position within the established order, shifting from point to point, like transposing scalar values in Euclidean space (Isin, 2002). Not surprisingly, therefore, premodernity did not privilege individual agency, but rather deference and compliance. Granted, some scholars explored alternative conceptions, but as the trial of Socrates illustrates, encouraging autonomous critical thought could be a crime punishable by death (Hackforth, 1972). Agentic performance was assessed in terms of adherence to norms, and the purpose of feedback was to refine replication and correct deviation. Thomas à Kempis (1952) epitomized this perspective in *The Imitation of Christ*, one of the most revered texts of the premodern Christian period. He explains that fulfillment comes from absorbing scripture and imitating the life of Christ, not from autonomous reasoning and choice. The latter perspective had to wait for Martin Luther, who nailed his 95 theses to the door almost a century later. In summary, premodern, agentic metamodels prioritized replication, rather than adaptive change or original composition.

Replicative Metamodels of Agency

Figure 2.1 depicts a premodern, replicative metamodel of agency. It builds on the cognitive-affective model of personality developed by Mischel and Shoda (1998). The figure assumes an input-process-output view of personality and the self, situated in context, but with relatively low variability and high stability. Hence, the metamodel illustrates a "persons in context" perspective, albeit with relatively low levels of contextual and system variation. Given these assumptions, the figure includes a sequence of major phases. First, it shows situational input stimuli (labeled SI), which include information about situations and problems in the world. Second, these stimulate sensory perception (SP), which transmits information to the next stage, cognitive-affective processing (CA). Third, the agent then processes information using cognitive-affective processing units (PU), which may include encodings, beliefs, affective states, goals, and values, including reference criteria and core commitments (RC), and self-regulatory schemes. Fourth, the agent generates action plans (AG), which are self-directed and self-regulated, to some degree. Fifth, these plans result in behavioral-performative outputs (BP). Sixth, such outputs trigger evaluation of performance (EP), as agents compare outcomes to aspirations and expectations, conditional on their degree of sensitivity to variance.

The figure also depicts three mechanisms of update encoding which flow from the evaluation of performance. These are shown by the arrowed

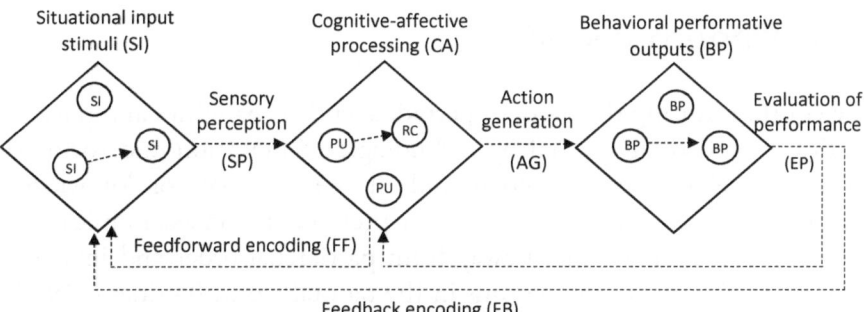

Fig. 2.1 Replicative agentic metamodel

lines at the base of the figure. The first two are feedback mechanisms (FB) which flow from evaluation of performance (EP). One is an inter-cyclical mechanism which updates cognitive-affective processes; that is, updates occur at the completion of full cycles of processing. A second inter-cyclical mechanism updates input stimuli and the situational context itself. In addition, there is an intra-cyclical feedforward mechanism (FF), flowing from cognitive-affective processing, which updates the situation itself and input processing. It is intra-cyclical, in relative terms, because it occurs during and influences the ongoing process. Natural, human feed-forward guidance is neurological and largely unconscious (Basso & Belardinelli, 2006), and was inherently part of human functioning, even in premodern times.

Furthermore, each phase is composed of functional components, illustrated by small circles. These are situational inputs (SI), cognitive-affective processing units (PU), including one dotted circle showing a referential criterion or commitment (RC), and behavior performances (BP) (see Mischel & Shoda, 1995). In the replicative metamodel, many components are invariant owing to stable contexts and behavioral norms. Arrows are dashed, to indicate relatively low variance and potentiality. Similarly, the figure depicts weak mechanisms of feedback and feedforward encoding, also shown by dashed lines. Premodern cultures did not exhibit widespread self-reflective variation, in this regard, but rather compliant imitation. Equally, learning was largely a process of memorization. As noted above, agents were guided by replication and imitative processes.

The Modern Period

By contrast, during the modern period, agentic capabilities and potentialities expanded, supported by technological innovation and socioeconomic development. For good and ill, new sources of knowledge, production, and mobility disrupted the premodern socioeconomic order. The focus of agency shifted away from patriarchal order and imagined beings, toward reasoning persons in the natural world (Giddens, 1991). Identity and meaning were now contingent on learning and achievement, rather than compliant acceptance of inherited position. That said, docility within social collectives remained hugely important, but it now became a political and philosophical question, versus one of theological dogma (e.g., Locke, 1967). In this fashion, modern criteria of reality, truth, and reasoning transcend premodern replication. Furthermore, the nature of human empathy and commitment frame modern thought about justice and ethics, rather than divine personalities and their pronouncements.

Consequently, the development of intelligent capabilities and the provision of opportunities for personal development and learning have been central to modern human science and theories of agency. Scholars examined the functioning of mind and personality, and the potential impact of social and cultural forces on human development (Pinker, 2010). Educational and clinical interventions built on such research. In parallel, the modern sciences became deeply dualistic. As noted in Chap. 1, for many scholars, human mind and consciousness were distinguished from material nature and the body. Hence, the human and natural sciences bifurcated into separate systems of study, with different methods of observation and analysis. Most critically, while the fundamental realities of mind and self may be directly accessible to ordinary consciousness and intuition, understanding of the natural world demands specialist technologies in controlled settings.

Reaction to Darwin's theory of evolution epitomizes this divide (see Mayr, 2002). On the one hand, the biological world was reconceived as a fully natural system, requiring scientific methods of observation and analysis, not driven by essentialism or teleology. On the other hand, however, many continued to believe that the fundamental features of mind and self were accessible to ordinary consciousness and irreducible to natural cause. Indeed, they feared that natural mechanisms would erode the ontological status of self-consciousness, and with it, various precepts of identity and faith. Subsequent debates reflect this dualism of modern thought: how to reconcile and integrate material cause and natural evolutionary mechanisms, with human consciousness, intentional action, and the interpretation of meaning.

Modern agentic metamodels exhibit the same tension. Most are deeply dualistic and assume problematic relations between the material and conscious aspects of human experience, or in other words, between mind and body. Reflecting this dualism, the major problems of modern agency can be compared to opposing vectors in Cartesian space: material versus intentional cause; natural selection versus preferential choice; biological instinct versus autonomous will (Reill, 2005). These polarities combine and often clash, in metamodels of evolutionary change and adaptive learning. Modern human science then seeks to resolve the resulting dilemmas. It asks, how do biological evolution and development interact with conscious mind and learning? Nevertheless, both mental and material processes involve change and development, albeit via different mechanisms. In consequence, the dominant agentic metamodels of modernity are broadly adaptive, rather than replicative.

Modern Adaptive Metamodels

Modern adaptive metamodels of agency therefore assume autonomous, reasoned problem-solving, learning, and development, within natural and cultural worlds. Persons are generally described as complex, open, adaptive systems, embedded in context (Shoda et al., 2002). Figure 2.2 illustrates this type of adaptive metamodel. Once again, it includes the same broad phases: situational input stimuli (SI) trigger sensory perception (SP), which in turn stimulate cognitive-affective processes (CA) by interacting processing units (PU), including one dotted circle showing a referential criterion or commitment (RC). These processes lead to action generation (AG) and resulting behavioral-performative outputs (BP), and the evaluation of performance (EP), which results in feedforward and feedback encoding (FF and FB respectively). Importantly, the modern, adaptive metamodel assumes stronger capabilities and more advanced technologies, when compared to the premodern, replicative metamodel in Fig. 2.1.

The figure depicts other important changes, compared to the replicative metamodel. To begin with, the metamodel in Fig. 2.2 is more complex, shown by additional component circles in each segment. The system is also more connected and dynamic, shown by the greater number of arrows, which are now solid rather than dashed, indicating stronger capabilities and potentialities. Hence, Fig. 2.2 shows greater functional intensity overall. Particularly, cognitive-affective processing is a more complex

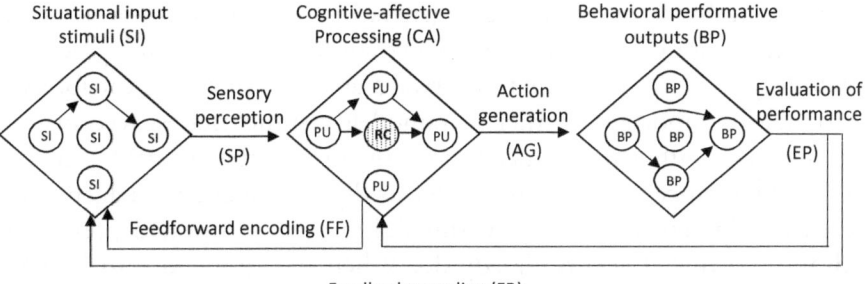

Fig. 2.2   Adaptive agentic metamodel

system of interacting units. Some of these units—especially beliefs and values—will also serve as reference criteria and core commitments, which guide action and the evaluation of outcomes. One such criterion or commitment is depicted by a dotted circle (RC). In actual systems, there will be many.

In addition, the adaptive metamodel in Fig. 2.2 has stronger feedback and feedforward mechanisms, indicated by the solid arrowed lines at the base of the figure. These solid lines represent the fact that variance often triggers adaptive learning in modern contexts, as well as updates to the stimulus environment and the system itself. Modern agents therefore exhibit stronger reflexive functioning, compared to agents in premodernity. Inter-cyclical feedback (FB) is more active as well. Moreover, some updates will amend reference criteria and core commitments. Although, scholars continue to debate which reference criteria and commitments are adaptive, when, why, and to what degree.

The Period of Digital Augmentation

In response to digitalization, agentic metamodels are transforming again. The technological assistance of agency is transitioning to a new level of scale, speed, and sophistication, thus driving a qualitative shift in agentic capability and potentiality. To begin with, recall that advanced artificial agents can compose, decompose, and recompose metamodels, in a dynamic fashion, potentially self-generating without external supervision. In addition, they learn with extraordinary speed and precision, including from intra-cyclical feedforward updates. This latter capability is particularly important. Earlier metamodels assume modest feedforward mechanisms, either from unconscious instinct or the effortful guidance of complex processes over time, whereas digitally augmented agents will learn rapidly and constantly in this fashion. As mentioned previously, advanced artificial agents already do. When incorporated into human-artificial collaboration, therefore, rapid feedforward learning, plus the sheer power and reach of artificial agency, transform agentic functioning. Augmented agency is intensely generative and near composability. Indeed, these capabilities distinguish digitalization from earlier periods of technologically assisted agency.

All aspects of processing are affected. Augmented agents can sense and sample the world more extensively, organize and process vast amounts of information very rapidly, represent and solve complex problems, and then design and direct responsive action. Augmented agents also achieve unprecedented speed and precision in learning, compared to purely human agents. To illustrate, consider the artificial agency required for autonomous mobility systems: constant real-time sensing of the environment, vehicles, and passengers; rapid complex problem-solving and empathetic interaction; accurate and coordinated action planning and control. Experts therefore apply a framework for autonomous vehicles known as "sense, plan and act" (Shalev-Shwartz et al., 2017). These systems exemplify that generative, augmented agency constitutes a new metamodel of intelligent agency (see Caro et al., 2014). Its central characteristics include the following: close collaboration between human and artificial agents; high sensitivity to context; intelligent sampling, representation, and resolution of complex problems; compositive methods, based in artificial intelligence; high sensitivity to variance and rapid evaluation of performance; very rapid processing and learning rates; real-time monitoring, self-regulation, and adjustment; often self-generative with minimal external supervision.

This historic transformation has meaningful topographical analogues in formal modeling. First, digital augmentation far transcends the state changes of the premodern period, which are comparable to scalar transpositions, shifting from point to point in Euclidean space. Second, digitalization also transcends the adaptive learning of modernity, which can be mapped as vector transitions in Cartesian space. Now third, and in contrast to both earlier periods, digital augmentation is about generative composition, which can be expressed as multi-vector tensor transformations, curving through Riemannian space (Kaul & Lall, 2019).

Generative Metamodels

Figure 2.3 illustrates this kind of generative metamodel of agency. Once again, the system integrates situational input stimuli (SI) which trigger sensory perception (SP), which in turn stimulate cognitive-affective processes (CA) consisting of interacting processing units (PU), again including one dotted circle showing a referential criterion or commitment (RC). As before, processing results in action generation (AG) leading to behavioral-performative outputs (BP), and the evaluation of performance processing and outcomes (EP), which often results in feedforward and feedback encoding (FF and FB respectively). However, this metamodel is more complex and dynamic, compared to the metamodels depicted in Figs. 2.1 and 2.2. This is shown by the greater number of component

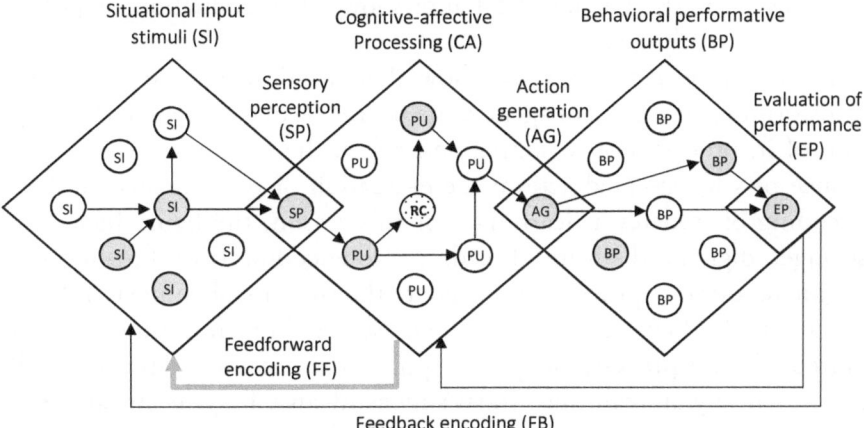

Fig. 2.3   Generative agentic metamodel

circles within each segment. Some are now shaded as well, which indicates they are digitalized, transformed by the incorporation of digital technology. Importantly, the major phases shown by large diamond shapes now partially overlap, integrated by digitalized processes.

As Fig. 2.3 further shows, digitalization occurs throughout the metamodel. First, the generation of situational inputs is increasingly digitalized. For example, through the internet of things and intelligent sensors, situational contexts are increasingly digitalized and connected. In consequence of this development, the figure also shows that sensory perception overlaps with cognitive-affective processing. Both phases are digitally intermediated. Second, cognitive-affective processing is equally digitalized and collaborative, whereby artificial agents interact with the cognitive-affective system. In this regard, recent innovations include cognitive computing, wearable devices, and artificial personality (Mehta et al., 2019). Reflecting this development, the figure shows that action generation is digitalized and overlaps cognitive-affective processing and behavioral-performative outputs. That is, digitalized action generation intermediates cognitive-affective processing and behavior performance. Current examples of this include artificial assistants and expert decision support systems (Wykowska, 2021). Third, behavioral-performative outputs are themselves digitalized, for example, by the incorporation of artificial agents, collaborative robotics, and intelligent prosthetics (Vilela & Hochberg, 2020). Finally, the figure shows that evaluation of performance is partially digitalized too.

In summary, Fig. 2.3 shows how digitalization is augmenting and transforming all aspects of agency: the stimulus environment and perception of it; processes of reasoning and affect; the generation and performance of self-regulated action; the evaluation of performance, and the encoding of updates as learning. For this reason, the figure includes a stronger, digitalized stream of feedforward encoding (FF), flowing from cognitive-affective processing to update the situational context and processing itself. Now depicted by a heavier shaded line, indicating it is digitalized. Via this process, the system updates the context and process itself, intra-cyclically, in real time. Today's most advanced agents already function in this way. Newer technologies, including devices which integrate real-time biometric feedback and augmented reality, will accelerate this trend. Feedforward updating will be constant and ubiquitous. In consequence, augmented agents will be increasingly self-generative, far eclipsing the agentic potentiality of earlier periods. These will be distinguishing features of generative, agentic metamodels.

At the same time, however, owing to their complexity and dynamics, these metamodels will be more difficult to supervise. Artificial and human agents will interact in every phase and function. However, as the previous chapter explains, both agents function in different ways. Much human processing is relatively myopic, sluggish, layered, and approximating, while artificial agents are increasingly fast, expansive, compressed, and precise. In consequence, many artificial feedforward mechanisms are inaccessible to human consciousness, and hence the two levels of processing could easily diverge. If this happens, augmented agents risk dysfunctional combinations of precision and approximation, fast and slow processing rates, sensitivity and insensitivity to variance, layering and compression, and complexity plus simplification. Artificial agency could then outrun, overwhelm, and bypass human inputs. Alternatively, human myopia and bias may infect artificial agents, and digitalization would then reinforce and amplify the limitations of human functioning. In summary, digitally augmented, generative metamodels pose major supervisory challenges.

2.2   Agentic Activation Mechanisms

Agentic activation mechanisms are being digitally transformed as well. To begin with, consider Fig. 2.2 once again, which shows the modern, adaptive metamodel of agency. All components are clearly distinguished and bounded. They are exogenous (external) or endogenous (internal), relative to each other. For example, situational inputs (SI) are exogenous to cognitive-affective processing (CA), whereas processing units (PU) are endogenous to cognitive-affective processing (CA). Now compare the digitally augmented, generative metamodel in Fig. 2.3. The figure is more highly integrated, with digitalized components connecting the major stages of situational inputs (SI), cognitive-affective processing (CA), and behavior performances (BP). These stages now overlap, owing to the digitalization of activation and intra-cyclical feedforward mechanisms (FF). Via such means, augmented agents will update the system in real time (Ojha et al., 2017). These mechanisms are central to the generative metamodel of agency.

Regarding the first two stages, digitalized mechanisms of sensory perception (SP), join the two large diamond shapes of situational inputs (SI) and cognitive-affective processes (CA). As a result, sensory perception becomes an intelligent process itself, thanks to the rapid intra-cyclical management of attention and sampling (see Fiedler & Wanke, 2009). In fact, environmental sampling and data gathering become deliberate, intelligent, and adaptive activities. Recent evidence supports this shift (e.g., Dong et al., 2020). The internet of things, smart sensors, wearable devices, and automated systems of multiple kinds, all connected to artificial agents, enable intelligent sensing and sampling, which radically complement ordinary sensory perception. However, from a modeling perspective, the digitalized mediation of intelligent sensory perception (SP) is neither exogenous nor endogenous, with respect to situational inputs (SI) and cognitive-affective processing (CA). Rather, intelligent sensory perception is in-between, mediating the boundaries of both the situational context and the cognitive-affective system.

Second, in like fashion, Fig. 2.3 shows that action generation (AG) is becoming behavioral and performative, not simply an antecedent of behavior. Action generation is now digitalized and joins cognitive-affective processing (CA) and behavior performance (BP). This means that action plans can be updated and regenerated during performances, in real time, via intra-cyclical feedforward mechanisms (Heaven, 2020). Artificial agents will process rapidly in the background, to integrate and update each phase of the process. Performances thus become more dynamic, thanks to digitalization. Hence, we can refer to performative action generation in generative metamodels. In fact, this already occurs in the development of agile, self-correcting systems (Howell, 2019). However, performative action generation (AG) is neither exogenous nor endogenous, with respect to cognitive-affective processing (CA) and behavior performance (BP). Rather, the process is again in-between, mediating the boundaries of both the cognitive-affective system and behavior performance.

Third, Fig. 2.3 shows that situational updating from feedback (FB) and feedforward (FF) is becoming intelligent itself. That is, situational contexts are becoming sites of intelligent learning, not simply passive sources of sensory inputs and problems. Once again, enabling technologies include the internet of things, ambient computing, and autonomous agents. All are embedded into the problem context and capable of updating it, often autonomously, in a self-generative fashion. Via these mechanisms, contexts will update and regenerate during problem-solving, not only from inter-cyclical, adaptive feedback. Hence, existing situations will evolve, and new ones emerge, during problem-solving itself. I describe this process as contextual learning, which is neither exogenous nor endogenous, with respect to behavior performance and the stimulus environment. Rather, the process is also in-between, mediating the boundaries of both the evaluation of performance (EP) and situational inputs (SI).

Augmented In-Betweenness

All the mediating mechanisms just described are central to augmented agency: intelligent sensory perception, performative action generation, and contextual learning. They are novel and transforming. Together, they allow augmented agents to learn, compose, and recompose in a dynamic fashion, updating form and function in real time. Augmented agency is therefore near composability, not only near decomposability. Moreover, these mechanisms signal a wider shift, from fixed boundaries and categories to fluid metamodeling. In consequence, however, being inside or outside of system boundaries at any time (endogenous or exogenous respectively) is often ambiguous and may not apply. This is because digitalized mediators operate at a higher rate and level of sensitivity, monitoring and adjusting system boundaries (see Baldwin, 2018). They are neither endogenous nor exogenous, relative to the boundaries they help to define. Rather, they are consistently in-between, processing potential form and function. These mechanisms will be critical and ubiquitous within digitally augmented agents.

Over recent years, scholars pay increasing attention to such effects. This interest is captured by the growing number of studies about forms of

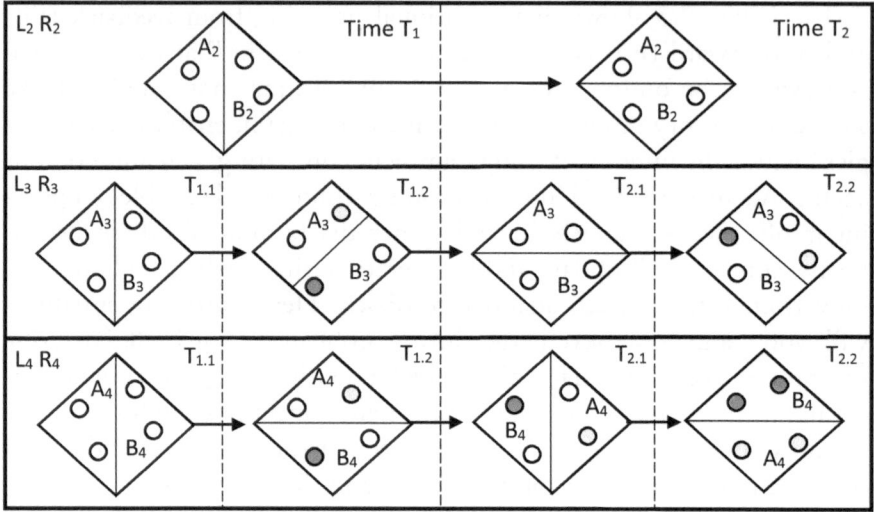

Fig. 2.4 Endogenous, exogenous, and in-between

ambiguous meaning, ambivalent value and belief, organizational hybridity, and ambidextrous action (e.g., March, 2010; O'Reilly & Tushman, 2013). In many domains, agents combine alternative, complementary, and sometimes conflicting patterns of thought and action, as they grapple with increasing phenomenal complexity and dynamism. This leads to shifting categorical boundaries, forms, and functions. The prefix "ambi," meaning both in Latin, is therefore a recurrent prefix in descriptive terms. As the reader will see in subsequent chapters, I exploit this prefix to describe other, novel patterns of in-betweenness arising from digitalization.

Figure 2.4 illustrates this type of dynamic mediation. It shows three levels and rates of processing and highlights the potential for divergence within augmented agents. To begin with, the upper third of the figure is modern adaptive capabilities at level L_2. They cycle at rate R_2 over times T_1 and T_2, and process inter-cyclical feedback at T_2. The middle third of the figure is stronger digitalized capabilities at level L_3 which cycle at rate R_3 over times $T_{1.1}$, $T_{1.2}$, $T_{2.1}$, and $T_{2.2}$. In other words, digitalized processing L_3R_3 cycles more rapidly, compared to the modern adaptive scenario L_2R_2. The lower third of the figure illustrates digitalized capabilities at level L_4

with cycle at rate R_4. These are roughly equal to L_3R_3 but are labeled differently to distinguish them. All three processes then depict the same two subsystems labeled A and B and their components. For example, these could be complementary subsystems of a problem-solving process.

However, the three levels of processing produce different patterns, shown by shaded dots and shifting boundaries of A and B. First, consider the upper portion of the figure, depicting L_2R_2. It shows that some of the components of A_2 and B_2 transition between times T_1 and T_2, and hence the boundary line shifts from being horizontal to vertical. For example, perhaps some components of solution search at T_1 become aspects of problem representation at T_2, reflecting adaptive learning which improves attention and problem sampling (Fiedler & Juslin, 2006). Now compare the middle process at L_3R_3. Components and boundaries shift as well, but in a different fashion. Most notably, the process cycles more rapidly, and as a result, the system changes at time $T_{1.2}$. One component of each subsystem has moved, along with the system boundary. The light gray dot shows a component at $T_{1.2}$ which is endogenous to A_3 and exogenous to B_3, but it remains endogenous to B_2. While the dark gray dot shows a component at $T_{1.2}$ which is endogenous to B_3 and exogenous to A_3, but it remains endogenous to A_2. Importantly, if we now combine the two subsystems (L_2R_2 and L_3R_3) in one augmented agent, some components are regularly in-between, simultaneously endogenous and exogenous, depending on the level of processing.

Note that at time T_2, both L_2R_2 and L_3R_3 are equivalent again. This could mean that the change at $T_{1.2}$ has now been incorporated into the system L_2R_2 via adaptive feedback. The same pattern of processing then occurs over the following cycle T_2. Once again, some shaded components are in-between, simultaneously endogenous and exogenous, relative to different levels of processing. But the degree of divergence is modest, only one component of each subsystem at a time. Moreover, L_2R_2 and L_3R_3 will synchronize at the completion of each major cycle; although in-between, they exhibit ambiguous boundary conditions. In summary, this augmented agent is broadly convergent over time, because digitalized intra-cyclical, feedforward updates at L_3R_3 are incorporated into L_2R_2 via inter-cyclical adaptive feedback. The agent absorbs updates effectively at both levels. Learning is generative and functional, in these respects.

Next, consider the third process at L_4R_4. Once again, components and boundaries shift, but now more extensively, compared to the other processes. At time $T_{1.2}$ components of A_4 and B_4 have moved, and the boundary between them is now vertical. Once again, the light gray dot shows a component at $T_{1.2}$ which is endogenous to A_4, but it remains endogenous to B_2. While the dark gray dot at $T_{1.2}$ is endogenous to B_4, but it remains endogenous to A_2. Moreover, if we combine the two subsystems (L_2R_2 and L_4R_4) in one augmented agent, half the components are in-between, simultaneously endogenous and exogenous, relative to different levels of processing.

Furthermore, at time T_2, the two processes L_2R_2 and L_4R_4 remain divergent, in contrast to the earlier convergent condition. Shaded components are persistently in-between, ambiguously endogenous and exogenous. In other words, the change in L_4R_4 at $T_{1.2}$ is not fully incorporated into the system L_2R_2, probably because the degree of digitalized processing at L_4R_4 is beyond the absorptive capabilities of L_2R_2. Moreover, the same pattern of divergent processing occurs again over the following cycle T_2. It leads to a compounding effect. By time $T_{2.2}$, all components of A and B are ambiguously endogenous and exogenous. This augmented agent is therefore increasingly divergent over time because digitalized intra-cyclical, feedforward updates at L_4R_4 are not incorporated into L_2R_2 via inter-cyclical adaptive feedback. The agent does not absorb updates effectively across levels and modalities. Learning is digitalized but dysfunctional, in these respects.

In fact, actual systems already exhibit these effects (e.g., Lee & Ro, 2015), for example, in the dynamic adaptation of modular transaction networks and software architecture (Baldwin, 2008). These processes rapidly update modular components, setting and resetting system boundaries. However, challenges escalate when inter-cyclical feedforward processing is fast and constant. Boundaries are consistently in flux. Hence, in highly digitalized systems, there will always be some components which are in-between, ambiguously endogenous and exogenous. The risk is that rapid, intra-cyclical updates will lack coordination with slower, inter-cyclical feedback, as depicted in Fig. 2.4. When this occurs in augmented agents, artificial and human processes will diverge, possibly leading to dysfunctional outcomes. Effective supervision will be critical.

Entrogenous In-Betweenness

Standard concepts fail to capture these novel features of digitalization. Indeed, most human sciences view in-betweenness as transitional, temporary, or paradoxical. The closest concept is liminality, but even it implies being ephemeral, and permanent liminality is viewed as dysfunctional, a sign of faulty, incomplete processing (Ibarra & Obodaru, 2016). Therefore, to capture this novel type of ongoing in-betweenness, another term will be helpful. I propose "entrogenous" which builds on "entre," meaning between in numerous European languages. Applied to the generative metamodel of agency, "entrogenous" and "entrogeneity" refer to the digitalized mechanisms which mediate in-betweenness, and whereby forms and functions develop and transform. Notably, such mechanisms are neither endogenous nor exogenous, relative to fixed boundaries. Rather, they are constantly in-between and mediating potential boundaries. The major risk, as shown by Fig. 2.4, is that poor supervision of entrogenous mechanisms will lead to divergent processes and dysfunctional outcomes. This particularly applies to the novel, digitalized mediators of augmented agency identified earlier: intelligent sensory perception, performative action generation, and contextual learning.

In fact, this puzzle is far from new. In ancient Greece, Heraclitus famously wrote, "You cannot step into the same river twice, for other waters are continually flowing on." Equally important but less well known, he also wrote, "We step and do not step in the same rivers. We are and are not" (Kirk, 1954). In other words, human experience, thought, and action are inherently in-between, constantly in flux. Form and function are relative to the frame of reference. As Heraclitus observed, a river is defined by its banks and flowing waters, and therefore simultaneously stable and always changing. In-betweenness is then normal, not dysfunctional. This Heraclitian perspective contrasted the thought of Plato and Aristotle, who favored categorical stability and essential order. Digitalized agentic systems address this ancient dilemma, because they allow for continual composition and recomposition at multiple levels. Hyperparameters and parameters are set and reset, as processes unfold (Feurer & Hutter, 2019). With respect to augmented agency, form and function will stabilize for a time, depending on the context, and recompose as contexts change. This is the core dynamic of generative, augmented agency. By analogy, therefore, entrogenous mediation is Heraclitian rather than Aristotelian. The implication being, that any instance of perceived permanence is mediated by some process in constant flux. To cite Herbert Simon (1996) again, science often transforms assumed states into dynamic processes.

At the same time, human and artificial agents possess different, inherent capabilities and potentialities. Much human processing is relatively sluggish, parochial, and heuristic, while artificial agents are increasingly fast, expansive, and precise. Therefore, what is entrogenous for artificial agents, may appear exogenous or endogenous for humans. As Heraclitus wrote, "We are and are not." These entrogenous dilemmas amplify the risks identified previously. Human agents could import inflexible categories, beliefs, and biases into augmented agency. Endogeneity and exogeneity would be baked in, from a human perspective. At the same time, however, artificial agents could relax categorical boundaries, and allow for greater plasticity and variation. The overall result will be divergent, and potentially conflicting, agentic form and function. In fact, these dilemmas are already observed in semi-supervised, collaborative systems (Kouvaris et al., 2015). They will be even more pronounced in larger, augmented communities and organizations. Later chapters will revisit this issue in relation to specific functional domains.

Summary of Metamodels

We can now summarize the foregoing discussion. To begin with, there are major differences between human and artificial agents. Natural human agents tend toward stability and possess limited capabilities and potentialities. As a result, purely human metamodels are relatively stable, weakly assisted by technologies, and not highly adaptive. They often possess deeply encoded hyperparameters, parameters, and variables, and fixed boundary conditions. Figure 2.1 illustrates this type of system labeled the replicative, agentic metamodel, which was dominant during premodernity. Next, as capabilities and technological assistance advanced, human agents became more autonomous and developmental, as shown in the modern, adaptive metamodel in Fig. 2.2. Granted, significant limits remain. Nevertheless, the adaptive metamodel of modernity affords greater degrees of freedom, compared to replicative metamodels. Boundaries are more adaptive and less fixed.

Digital augmentation now promises far greater capabilities and potentialities. Most notably, digitalization enables augmented agents which combine human and artificial agents in close collaboration. A major feature will be the capability for generative metamodeling, effectively in real time. Digitalized entrogeneity will be fundamental, mediating the dynamic composition of agentic form and functioning. This type of digitally augmented, generative metamodel is shown in Fig. 2.3. If well supervised, augmented agency and humanity will enjoy greater degrees of freedom and potentiality. But there are also major risks and dilemmas to resolve.

2.3   Dilemmas of Digital Augmentation

The greatest benefits of digital augmentation are its potential weaknesses. As often happens, remarkable strengths easily skew performance outcomes. On the one hand, augmented agents will sample and search ever more widely, process information at increasing speed, scale, and accuracy, and learn at unprecedented rates. On the other hand, thanks to human semi-supervision, augmented agents will often inherit myopias, biases, and parochial commitments. Therefore, digitally augmented agency confronts a fundamental challenge: how to combine and supervise human and artificial capabilities while avoiding excessive divergence, convergence, and distortion? Resolving these questions will be critical for augmented humanity.

Problematics and Metamodels

To clarify these topics further, Fig. 2.5 summarizes the agentic metamodels and problematics already discussed. It shows three agents X, Y, and Z, over three successive time periods labeled 1, 2, and 3. The figure captures the essence of three broad historical periods, premodernity, modernity, and contemporary digitalization. First, recall that in premodern contexts, metamodels of agency assume low complexity, relatively poor capabilities and potentialities, with little variation. Agency tends to be imitative and replicative. Overall agentic functioning is viewed as a collective accomplishment, rather than an outcome of autonomous individuals. Predictably, therefore, premodern problematics focus on the integration of persons within communal narratives, and how to account for variation in a world of ordained stability and order (Walker, 2000). In Fig. 2.5, these conditions are shown by the segments with only one dot in each, which exclude agent X at time 3. Assume that each dot represents components of some agentic function. As the figure shows, functioning is widely dispersed across the collective (agents X, Y, and Z) over time (periods 1, 2, and 3). Each individual agent is weakly responsible for overall functioning at any time, apart from agent X at time 3. Hence, they are highly dependent on each other. Therefore, if we assume that all the segments with one dot are required to perform a particular agentic process, then efficacious action will require the cooperation of all three agents over the three time periods, and hence the minimization of individual variance. In this way, the segments with one dot expose core features of the premodern, replicative metamodel and its associated problematics. Individuals must conform and cooperate over time, to achieve collective outcomes.

Notably, simple models of artificial intelligence and machine learning possess similar features. They, too, reference encoded models to solve predictable types of problems (Norvig & Russell, 2010). Moreover, these simpler model-based, artificial agents, are frequently embedded within processing networks, just like the agents in Fig. 2.5 with only one major functional role. Functioning is therefore highly distributed. In these respects, simpler types of artificial agent exhibit metamodels which are comparable to those of premodern agency. Such technologies are therefore less genuinely "agentic" and augmenting, and sit at the passive end of the supervisory spectrum. That said, this suggests a promising avenue for research into replicative metamodels. Simpler model-based agents appear well suited to the task.

By contrast, during modernity, human agents develop stronger capabilities for autonomous thought and action. Agentic functioning is both an individual and collective accomplishment, and numerous technological innovations assist these developments. Modern problematics therefore focus on the reconciliation of individual freedom and collective solidarity, and the means by which humans adapt and transcend their limited capabilities and potentialities (Pinker, 2018). These problematics are illustrated by agent X and time 3, excluding the more densely dotted segment at the base of the figure. Notably, there are now extra dots within agent X at time 3, which shows that this individual is more capable and performs more functions, thanks in part to increased technological assistance. Indeed, agent X contains as many functional components at time 3, as all other agents, which illustrates the agent's capability for autonomous action. Nevertheless, agent X remains reliant on the collective. Significant functions are still distributed, and effective performance will require cooperation with other agents over time. In these respects, agent X at time 3 illustrates core aspects of the adaptive metamodel and problematics of modernity, namely, how to develop individual capability and potentiality, while integrating with collective form and function?

Once again, there are strong parallels to artificial agents. In fact, the modern adaptive metamodel corresponds to goal-based and utility-based, artificial agents, which are more advanced than the simpler model-based agents discussed above (Norvig & Russell, 2010). First, goal-based artificial agents use encoded preferences to guide problem-solving. Other things being equal, they seek to achieve predetermined outcomes. Second, utility-based artificial agents possess additional rules for the rank ordering of potential outcomes and then seek to maximize utility. Clear parallels exist in modern social and behavioral theories. In many such disciplines, agents are conceived as goal seeking, maximizing preferences and utility (Bandura, 2007; Thaler, 2016). Hence, the architecture of goal-based and utility-based artificial agents, is broadly comparable to the adaptive metamodel of modernity. Both entail intelligent agents, working in concert, seeking to achieve goals and maximize preferences.

Contemporary digitalization supports a new, generative metamodel of agency. It assumes high levels of complexity, unprecedented processing capability, and intense patterns of functioning at every level. Figure 2.5 also illustrates the core features of such a metamodel, by showing one component of agent X at the base of time 3, which is very dense with

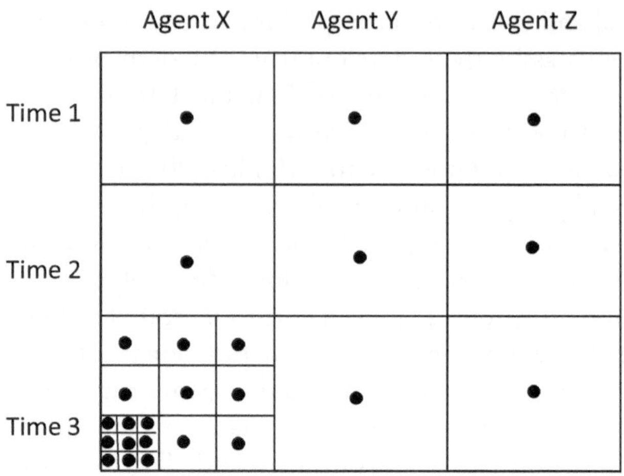

Fig. 2.5 Historical problematics of agency

dots. This component is fully digitalized. Indeed, this digitalized component of agent X exhibits as many functions as all other segments of X, as well as the other agents in the figure. In a fully digitalized collective, all agents and components will be equally intense. The increase in functional complexity is exponential, across multiple levels and modalities, and within any time period as well. New problematics thus emerge: how can human beings collaborate closely with artificial agents, while remaining genuinely autonomous, in reasoning, belief, and choice; how will human and artificial agents learn to understand, trust, and respect each other, despite their different capabilities and potentialities; and how will augmented agents supervise the dynamic composition and recomposition of metamodels? And not surprisingly, this generative metamodel mirrors the architecture of the most advanced artificial agents, because it assumes participation by such systems. Advanced artificial agents will be integral to augmented agency.

In summary, the different functional patterns in Fig. 2.5 capture the history of both artificial and human agency. The figure shows how the recent evolution of artificial agency shares important features with the long history of human agency, at least in terms of their metamodels. First, model-based artificial agents map to the replicative metamodels of premodernity. Second, goal-based and utility-based, artificial agents mirror the adaptive metamodels of modernity. And third, advanced artificial agents instantiate the generative metamodels of digital augmentation. In these respects, the rapid ontogeny of artificial agency over recent decades, recapitulates the slow phylogeny of civilized humanity over millennia (see Clune et al., 2012). Or to paraphrase Hegel (1980), the recent history of digital science recapitulates the digitalized science of history. More striking still, both processes converge in the science of augmented agency because the story of digital science mirrors the science of augmented humanity. Producing a historical synthesis Hegel would surely appreciate.

2.4   Patterns of Supervision

Even in a highly digitalized world, however, people will continue to exhibit models of agency which are effectively premodern, in terms of their core components, levels of complexity, and modes of supervision. There will be a spectrum of artificial augmentation. In some contexts, that is, agency will still be governed and supervised in terms of replication and narrative, as in many cultural and faith communities. Similarly, people will continue choosing modern adaptive metamodels which entail less intrusive technological assistance, as in many social and cultural pursuits. Therefore, earlier agentic options will remain feasible and often desirable, as in matters of faith and family. But they will exhibit reduced functionality, compared to fully digitalized, generative options. In fact, engineers plan for these options too, recognizing that people will sometimes wish to control technological functioning for recreational or other reasons (Simmler & Frischknecht, 2021). However, extra problems arise when agents adopt different metamodels at the same time. I will return to this topic in later sections.

In the meantime, as the preceding argument explains, a central feature of any agentic metamodel is the quality of its supervision. That is, how and to what degree, the metamodel is copied or composed, self-regulated or externally controlled, and from which source. We can therefore distinguish metamodels in terms of their supervision, and particularly, in terms of human and technological sources of supervision. Nine alternative patterns are depicted in Fig. 2.6. On the horizontal dimension, the figure shows the level of human supervision. While the vertical dimension shows the level of technological supervision. Each segment of Fig. 2.6 therefore shows two potential sources and three levels of supervision of agency, low, medium, and high. Circles with dashed borders represent technological supervision, while human supervision is represented by circles with solid borders. The circles in each segment overlap because both types of supervision interact. It is important to note, that the size of these shapes does not represent the absolute strength or complexity of supervision, but rather their relative significance in any metamodel.

First, consider segment 1 in Fig. 2.6. It shows the type of simple supervision in replicative metamodels of agency, which dominated during pre-modernity. Human supervision is shown by the small circle with a solid border, and technological supervision by the small circle with a dashed border. In this metamodel, therefore, technological and human levels of supervision are both low. Supervision is routine, encoded, and replicative, relying on communal rituals, perhaps simple tools for writing,

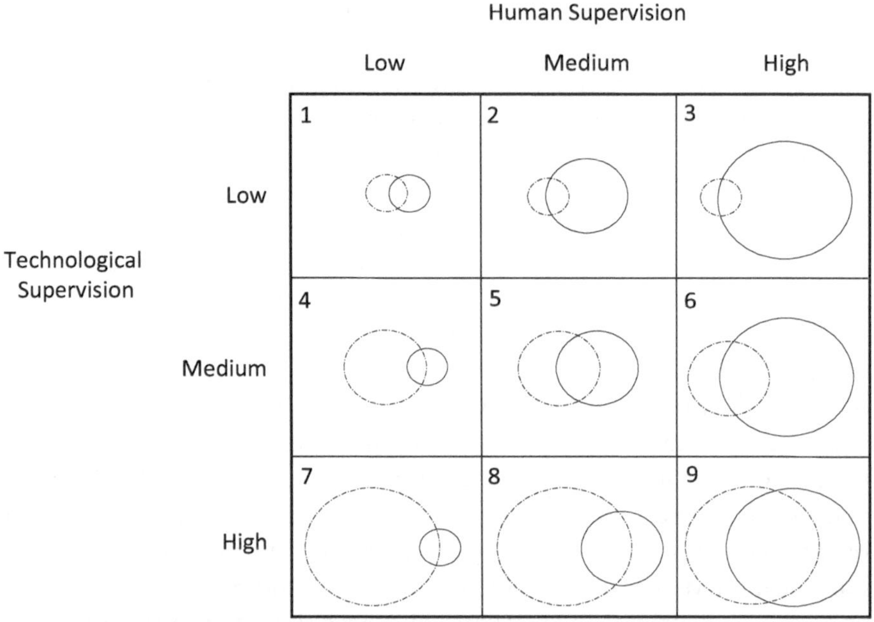

Fig. 2.6 Historical eras of agentic supervision

counting, and communicating, but not much more. In summary, replicative metamodels of agency have relatively low levels of autonomous, technological, and human supervision. The metamodels offer few degrees of freedom. In this regard, segment 1 complements the premodern problematics depicted earlier, by the segments in Fig. 2.5 which have only one functional dot.

Now consider segments 2, 4, and 5 as well. They show metamodels with greater technological capabilities, as in modernity. In fact, segments 1, 2, 4, and 5, represent the patterns of supervision in modern, adaptive metamodels of agency. They complement the earlier depiction of modern problematics in Fig. 2.5, and especially the role of agent X in the collective. Most notably, there are now four options in Fig. 2.6, combining medium or low human supervision, with medium or low technological supervision. In other words, both human and technological supervision have advanced. Relevant technologies are largely mechanical and analogue and provide a moderate assistance to the supervision of agency. Human capabilities also advance, at least for some people, but still within constraints. Indeed, the limits of human supervisory capability are a persistent theme of modernity. Hence, the metamodels of agency represented by segments 1, 2, 4, and 5 in Fig. 2.6 bestow greater degrees of freedom, compared to the preceding replicative option.

Regarding the details, segment 5 shows medium levels of human and technological supervision. This encompasses technologically assisted domains, such as surgical practice, in which human and technological supervision are both critical. Segment 2, on the other hand, shows dominant human supervision, similarly to premodern contexts. Segment 4 shows the opposite scenario, in which technological supervision dominates, as it often does in mechanical systems which operate independently of human intervention. Many automated processes are like this. In summary, modernity exhibits different levels of human and technological supervision, generating alternative agentic options. For this reason, modernity also presents more frequent choices and dilemmas, about which metamodel of agency fits best, when, and why.

Next, by including all 9 segments of Fig. 2.6, we have an illustration of digitalized, generative metamodels of agency. Clearly, there are more feasible metamodels and choices. Human capabilities are more developed in segments 3, 6, and 9, as are technological capabilities in segments 7, 8, and 9. Moreover, augmented agents can exploit all these metamodels, and potentially in real time. This dynamism will be significantly owing to the entrogenous mediation mechanisms discussed previously and illustrated in Fig. 2.4. But these mechanisms also bring new challenges and risks. If supervision is poor, agents might develop in overly divergent or convergent ways and become dysfunctional, for example, adopting the option in segment 3 when the balanced option in segment 5 is more appropriate. In this respect, the whole of Fig. 2.6 captures a central challenge for digitally augmented agents, namely, the complexity of supervising human-machine collaboration (see Murray et al., 2020; Simmler & Frischknecht, 2021).

Depending on the context, therefore, each metamodel in 2.6 can be effective and appropriate. To begin with, scenario 1 will be largely routine. Whereas segment 9 shows the opposite scenario. Human and technological supervision are both strong and assertive. This metamodel is best suited to highly digitalized, complex, dynamic contexts. Expert medical practice is a good example, in which artificial and human agents both supervise critical aspects of collaborative functioning. The major risk is conflict within the augmented agent, for example, when the circles in segment 9 overlap less and supervision is poorly coordinated. Other scenarios show the alternatives in-between, combining low, medium, and high levels of supervision. Sometimes human supervision is clearly dominant, as in scenario 3. This metamodel will be fully humanized and supervision will be guided by ordinary values and commitments. However, the risk is that human myopia and bias will intrude and distort the system. Next, there are metamodels in which technological supervision is dominant, as in scenario 7. These are highly digitalized, but the risk is that human inputs are excluded inappropriately. People could become digitally docile and overly dependent. Segment 5 is intermediate. It includes moderate supervision of both kinds, but significant freedom as well. This metamodel could be appropriate in exploratory, creative contexts. The risk is that agents may lack enough supervision and tend toward incoherence.

Furthermore, segment 1 in a digitalized world, has the same general pattern of supervision in a premodern context. In other words, the type of routine agency which dominated during premodernity may still occur within a digitalized world. Even in the period of digitalization, that is, people may adopt purely replicative models of agency, over digitally augmented options. This may seem counterintuitive, but in fact, it will be widespread. Earlier, I cited traditional cultural and faith traditions as examples of such choices. However, this might lead to the reinforcement of human myopias and biases. Next, segments 1, 2, 4, and 5 are equivalent in the adaptive and generative systems. What this means, is that the modern patterns of supervision can also occur in a digitalized world. People will still exhibit adaptive, modern approaches, and eschew high levels of digital augmentation. For example, purely human inputs and adaptive learning will likely remain dominant in some social and professional contexts. Moreover, such metamodels may be fully functional, assuming the choice of metamodel fits the context. The major challenge in all scenarios, therefore, is to develop mutual understanding, trust, and empathy, within the augmented agent.

2.5   Implications for Augmented Humanity

Significant dilemmas therefore confront digitally augmented humanity, conceived in terms of purposive agency, primarily because human and artificial agents have different capabilities and potentialities. Compared to artificial agents, humans are often myopic, sluggish, layered, and approximating. While relative to humans, artificial agents are increasingly expansive, fast, compressed, and precise. When combined in augmented collaboration, these divergent characteristics are either complementary or conflicting. If the collaboration is well supervised, they are complementary. Human and artificial agents strengthen each other and mitigate the other's limitations. However, if poorly supervised, they are conflicting, and combination leads to poorly fitting metamodels: either underfitting, meaning metamodels admit too much noise and variance, and fail to clarify potential models of interest; or overfitting, meaning they omit too much noise and potential variance, thereby excluding potential models of interest. Metamodels of agency can therefore skew inappropriately, either amplifying human priors, especially myopias and biases. Or they skew the other way, by amplifying digital processes which diminish and override the human. Sometimes both patterns of distortion will occur, resulting in extremely divergent systems. In each scenario, augmented agents will underperform and incur functional losses. These risks already attract significant attention from computer scientists (Gavrilov et al., 2018). Moving forward, they will be major concerns for human scientists as well. The supervision of augmented agency will require new theory and techniques.

Implications for Specific Domains

Hence, the following challenge arises for theory and practice: to understand which aspects of human and artificial supervision should be reinforced, adapted, or relaxed, so that augmented agents maximize the benefits of digitalization, while preserving core values, commitments, and other humanistic qualities. Important legacies are at stake. Many aspects of modernity, and even premodernity, could remain fulfilling and functional, even in a highly augmented world, assuming agents adopt these options appropriately and avoid superstitious thinking and distorting priors. For example, people will continue to find meaning in religious narrative and spiritual commitment, and purely human supervision could be fully appropriate in everyday life. Augmented humanity will benefit by preserving these options. The challenge is dynamically to determine which values and commitments are humanizing, and which are distorting, when and why, and then to supervise them effectively (Sen, 2018).

Collaborative supervision is therefore a major challenge for augmented humanity. It reflects the core problematic of digitalization: how to combine and balance human and artificial capabilities and potentialities? It will impact all aspects of agentic form and function. The following chapters examine important areas of impact. In doing so, they will emphasize different aspects of the generative agentic metamodel in Fig. 2.3:

- Chapter 3 is about agentic modality, which involves the overall metamodeling of individual, group, and collective form and function.
- Chapter 4 examines problem-solving, which involves sensory-perceptive (SP) inputs about problems, and cognitive-affective processing (CA) then searches for solutions.

- Chapter 5 examines cognitive empathy for other minds, which takes sensory-perceptive (SP) inputs about other minds, and cognitive-affective processing (CA) then searches for empathic solutions.
- Chapter 6 focuses on self-regulation, which is a major cognitive-affective processing unit (PU) and integral to action generation (AG), among other reflexive functions.
- Chapter 7 is about evaluation of performance (EP), which assesses the outcomes of behavior performances, referencing some criteria.
- Chapter 8 examines learning, especially in which evaluation of performance leads to feedforward (FF) and feedback (FB) updates.
- Chapter 9 is about self-generation, that is, how augmented agents choose and transform their own metamodels and narratives.
- Chapter 10 summarizes the argument and looks forward to a science of digitally augmented agency.

Reality and Truth

Additional consequences follow for the core criteria of reality, truth, and ethics, or in other words, for core ontological, epistemological, ethical, and cultural commitments. To begin with, recall that dominant agentic metamodels at any time, tend to reflect the limitations of capability and potentiality. From this perspective, ideal criteria are extrapolations of human limitation (Appiah, 2017). As an adaptive mechanism, such idealization is explicable and often functional. It bolsters agents' self-efficacy and sense of security because human capabilities appear to encompass the limits of reality, truth, and value. Ideals also provide predictable meaning and guidance. Of course, empirical science consistently exposes the contingency of such ideals, which partly explains why scientific advance is often controversial. It challenges inflated self-efficacy and identity. Ironically, therefore, the greater the impact of applied science, including digitalization, the more resistance it may provoke.

To illustrate, consider the contemporary period. In much modern thinking, material nature is opposed to conscious mind. Kant (1998) clearly established the distinction: pure abstract reasoning draws from ideal spirit and is categorically different from practical thought in the temporal realm. Rational mind was a category of immaterial reality, while practical reasoning translated mind into the contingent world. From a purely functional perspective, this belief liberates autonomous mind from premodern myth and superstition, while protecting it from mechanistic reduction. In addition, by separating mind and nature, it provides a rationale for human exploration and exploitation of the natural world. However, in the light of artificial intelligence and cognitive neuroscience, many now question these assumptions. They rightly observe that artificial intelligence blends cognitive and material phenomena, while neuroscience promises to explain consciousness in terms of natural and ecological mechanisms (Seth, 2018).

Furthermore, as Chap. 1 explains, the insights of digitally augmented neuroscience imply that the fundamental hyperparameters of mind and consciousness are not directly accessible to reflexive consciousness itself. Instead, researchers use neurophysiological and digital techniques. In fact, advanced artificial intelligence already simulates significant aspects of conscious mental life, including calculative and associative reasoning, intuition and increasingly, empathy and personality (Mehta et al., 2019). As these technologies mature, traditional distinctions between mind and nature, and between virtual and material, will appear increasingly contingent, more functional than fundamental. This shift is foreshadowed by the generative agentic metamodel in Fig. 2.3. It shows that digitalized processes infuse all areas of agentic functioning.

Value and Commitment

Earlier in this chapter, I also noted that ancient thought remains relevant today. For example, the ancients examined the hedonic nature of life, contrasting pleasure and pain, gain and loss, life and death. Not surprisingly, humans approach the former conditions and try to avoid the latter. These hedonic principles still run deep in Western thinking. To illustrate, Higgins' (1998) Regulatory Focus Theory is explicitly hedonic. It contrasts the prevention of pain and loss, against the promotion of pleasure and gain. As another example, Kahneman and Tversky's (2000) Prospect

Theory draws fundamental distinctions between potential gains and losses in behavioral decision-making. And to be sure, life was and is, deeply hedonic. Human beings naturally seek to avoid pain, loss, and death, while hoping for pleasure, gains, and life.

Going further, however, the ancients also thought about eudaimonic needs, or the human desire for overall well-being and to live a good life (Aristotle, 1980). These aspirations transcend the hedonic avoidance of loss and pursuit of gains. Rather, eudaimonia embraces the whole of experience, and hence, the totality of human purpose and potentiality (Di Fabio & Palazzeschi, 2015). Notably, this concept has gained fresh prominence in contemporary thought, including positive psychology (Seligman & Csikszentmihalyi, 2000), new thinking in economics by Amartya Sen (2004) and others, and value-based models of social organization (Lounsbury & Beckman, 2015).

This conceptual shift may be partly explained by the rapid growth of capabilities and potentialities brought about by modernity and its related changes. Put simply, flourishing is now more feasible for more people (Phelps, 2013). Of course, widespread discrimination, deprivation, and inequality persist, and new divisions have emerged. Nonetheless, owing in part to digital augmentation, the realm of agentic potentiality is expanding. Human beings will be able to curate new metamodels of being and becoming, including dynamic compositions of the self and community. Agentic potentiality will be vastly different in such a world. In fact, this transformation will likely spawn a new science of eudaimonics, integrating value commitments broadly conceived, where commitment in this context is defined as being dedicated, feeling obligated and bound, to some value, belief, or pattern of action. Such a science would complement the existing disciplines of economics, ethics, politics, and aesthetics (see Di Fabio & Palazzeschi, 2015; Sen, 2004). I will return to this possibility in the final chapter.

In the meantime, advances in cognitive neuroscience and computer science reinforce the fact that fundamental properties of mind and consciousness cannot be accessed via ordinary means (Seth, 2018). Introspection and intersubjectivity are no longer enough, nor the anthropomorphic conceptions which these methods support. New concepts and techniques are required. Yet at the same time, human beings live in and through ordinary consciousness. It is fundamental to being and remaining human. This presents a core challenge for augmented agency, which is to maintain the value and significance of consciousness and mental life, even as science frees itself from anthropomorphic constraints. In fact, this dilemma reinforces the role of commitments in the supervision of augmented agency, because commitments will anchor agents in lived experience. Commitments validate and sustain ordinary consciousness and mind, without claiming scientific status. They are simply and importantly human. Hence, commitments will play a central role in the supervision of augmented agency. They will reinforce humanistic values, helping to preserve the experience of ordinary mind and consciousness in a digitalized world.

References

Appiah, K. A. (2017). *As if: Idealization and ideals*. Harvard University Press.

Aristotle. (1980). *The Nicomachean Ethics* (D. Ross, Trans.). Oxford University Press.

Baldwin, C. Y. (2008). Where do transactions come from? Modularity, transactions, and the boundaries of firms. *Industrial & Corporate Change, 17*(1), 155–195.

Baldwin, C. Y. (2018). Design rules, volume 2: How technology shapes organizations; chapter 6 the value structure of technologies, part 1: Mapping functional relationships. *Harvard Business School, Harvard Business School Research Paper Series* (19-037).

Bandura, A. (2007). Reflections on an agentic theory of human behavior. *Tidsskrift-Norsk Psykologforening, 44*(8), 995.

Basso, D., & Belardinelli, M. O. (2006). The role of the feedforward paradigm in cognitive psychology. *Cognitive Processing, 7*(2), 73–88.

Caro, M. F., Josyula, D. P., Cox, M. T., & Jimenez, J. A. (2014). Design and validation of a metamodel for metacognition support in artificial intelligent systems. *Biologically Inspired Cognitive Architectures, 9*, 82–104.

Clune, J., Pennock, R. T., Ofria, C., & Lenski, R. E. (2012). Ontogeny tends to recapitulate phylogeny in digital organisms. *The American Naturalist, 180*(3), E54–E63.

Di Fabio, A., & Palazzeschi, L. (2015). Hedonic and eudaimonic well-being: The role of resilience beyond fluid intelligence and personality traits. *Frontiers in Psychology, 6*, 1367.

Dong, H., Ma, W., Wu, Y., Zhang, J., & Jiao, L. (2020). Self-supervised representation learning for remote sensing image change detection based on temporal prediction. *Remote Sensing, 12*(11), 1868.

Feurer, M., & Hutter, F. (2019). Hyperparameter optimization. In *Automated machine learning* (pp. 3–33). Springer.

Fiedler, K., & Juslin, P. (Eds.). (2006). *Information sampling and adaptive cognition*. Cambridge University Press.

Fiedler, K., & Wanke, M. (2009). The cognitive-ecological approach to rationality in social psychology. *Social Cognition, 27*(5), 699–732.

Gavrilov, A. D., Jordache, A., Vasdani, M., & Deng, J. (2018). Preventing model overfitting and underfitting in convolutional neural networks. *International Journal of Software Science and Computational Intelligence (IJSSCI), 10*(4), 19–28.

Geertz, C. (2001). *Available light*. Princeton University Press.

Giddens, A. (1991). *Modernity and self-identity: Self and society in the late modern age*. Stanford University Press.

Hackforth, R. (1972). *Plato: Phaedo*. Cambridge University Press.

Heaven, D. (2020, January 21). IBM's debating ai just got a lot closer to being a useful tool. *MIT Technology Review*.

Hegel, G. W. F. (1980). *Lectures on the philosophy of world history*. Cambridge University Press.

Higgins, E. T. (1998). Promotion and prevention: Regulatory focus as a motivational principle. *Advances in Experimental Social Psychology, 30*, 1–46.

Howell, B. M. (2019). Managing emerging technology and organizations with agility. In *Advances in the technology of managing people: Contemporary issues in business*. Emerald Publishing Limited.

Ibarra, H., & Obodaru, O. (2016). Betwixt and between identities: Liminal experience in contemporary careers. *Research in Organizational Behavior, 36*, 47–64.

Isin, E. F. (2002). *Being political: Genealogies of citizenship*. University of Minnesota Press.

Kahneman, D., & Tversky, A. (2000). Prospect theory: An analysis of decision under risk. In D. Kahneman & A. Tversky (Eds.), *Choices, values, and frames* (pp. 17–43). Cambridge University Press.

Kant, I. (1998). *Critique of pure reason* (P. Guyer & A. W. Wood, Trans.). Cambridge University Press.

Kaul, P., & Lall, B. (2019). Riemannian curvature of deep neural networks. *IEEE Transactions on Neural Networks and Learning Systems, 31*(4), 1410–1416.

Kempis, T. à. (1952). *The imitation of Christ (c. 1420)* (L. Sherley-Price, Trans.). London: Penguin Classics.

Kirk, G. S. (1954). *Heraclitus: The cosmic fragments.* Cambridge University Press.

Kouvaris, K., Clune, J., Kounios, L., Brede, M., & Watson, R. A. (2015). How evolution learns to generalise: Principles of under-fitting, over-fitting and induction in the evolution of developmental organisation. *arXiv preprint arXiv:1508.06854.*

Lee, S. H., & Ro, Y. M. (2015). Partial matching of facial expression sequence using over-complete transition dictionary for emotion recognition. *IEEE Transactions on Affective Computing, 7*(4), 389–408.

Locke, J. (1967). *Locke: Two treatises of government.* Cambridge University Press.

Lounsbury, M., & Beckman, C. M. (2015). Celebrating organization theory. *Journal of Management Studies, 52*(2), 288–308.

March, J. G. (2010). *The ambiguities of experience.* Cornell University Press.

Mayr, E. (2002). *What evolution is.* Weidenfeld & Nicolson.

Mehta, Y., Majumder, N., Gelbukh, A., & Cambria, E. (2019). Recent trends in deep learning based personality detection. *Artificial Intelligence Review*, 1–27.

Mischel, W., & Shoda, Y. (1995). A cognitive-affective system theory of personality: Reconceptualizing situations, dispositions, dynamics, and invariance in personality structure. *Psychological Review, 102*(2), 246–268.

Mischel, W., & Shoda, Y. (1998). Reconciling processing dynamics and personality dispositions. *Annual Review of Psychology, 49*(1), 229–258.

Murray, A., Rhymer, J., & Sirmon, D. G. (2020). Humans and technology: Forms of conjoined agency in organizations. *Academy of Management Review* (online).

Norvig, P., & Russell, S. (2010). *Artificial intelligence: A modern approach* (3rd ed.). Pearson.

O'Reilly, C. A., & Tushman, M. L. (2013). Organizational ambidexterity: Past, present, and future. *The Academy of Management Perspectives, 27*(4), 324–338.

Ojha, V. K., Abraham, A., & Snášel, V. (2017). Metaheuristic design of feedforward neural networks: A review of two decades of research. *Engineering Applications of Artificial Intelligence, 60*, 97–116.

Phelps, E. S. (2013). *Mass flourishing: How grassroots innovation created jobs, challenge, and change.* Princeton University Press.

Pinker, S. (2010). *The language instinct: How the mind creates language.* HarperCollins.

Pinker, S. (2018). *Enlightenment now: The case for reason, science, humanism, and progress.* Penguin.

Reill, P. H. (2005). *Vitalizing nature in the enlightenment.* University of California Press.

Seligman, M. E. P., & Csikszentmihalyi, M. (2000). Positive psychology: An introduction. *American Psychologist, 55*(1), 5–14.

Sen, A. (2004). Economic methodology: Heterogeneity and relevance. *Social Research, 71*(3), 583–614.

Sen, A. (2018). The importance of incompleteness. *International Journal of Economic Theory, 14*(1), 9–20.

Seth, A. K. (2018). Consciousness: The last 50 years (and the next). *Brain and Neuroscience Advances, 2*, 1–6.

Shalev-Shwartz, S., Shammah, S., & Shashua, A. (2017). On a formal model of safe and scalable self-driving cars. *arXiv preprint arXiv:1708.06374.*

Shoda, Y., LeeTiernan, S., & Mischel, W. (2002). Personality as a dynamical system: Emergence of stability and distinctiveness from intra- and interpersonal interactions. *Personality and Social Psychology Review, 6*(4), 316–325.

Simmler, M., & Frischknecht, R. (2021). A taxonomy of human–machine collaboration: Capturing automation and technical autonomy. *AI & SOCIETY, 36*(1), 239–250.

Simon, H. A. (1996). *The sciences of the artificial* (3rd ed.). The MIT Press.

Sorabji, R. (2006). *Self: Ancient and modern insights about individuality, life, and death.* OUP.

Spar, D. L. (2020). *Work mate marry love: How machines shape our human destiny.* Farrar, Straus and Giroux.

Thaler, R. H. (2016). Behavioral economics: Past, present, and future. *American Economic Review, 106*(7), 1577–1600.

Vilela, M., & Hochberg, L. R. (2020). Applications of brain-computer interfaces to the control of robotic and prosthetic arms. In *Handbook of clinical neurology* (Vol. 168, pp. 87–99). Elsevier.

Walker, J. (2000). *Rhetoric and poetics in antiquity.* Oxford University Press.

Williams, B. (1993). *Shame and necessity.* University of California Press.

Wykowska, A. (2021). Robots as mirrors of the human mind. *Current Directions in Psychological Science, 30*(1), 34–40.

3

Agentic Modality

Fortunately for humanity, many ecologies are stable and munificent over time. Civilization can flourish, notwithstanding episodic disasters and disruption. Social systems evolve and human beings cooperate in purposive action. These ecologies elicit and sustain different agentic modalities, or expressions of agentic form and function. Three such configurations consistently emerge: individual persons, relational groups, and larger collectives (Bandura, 2006). All three are interconnected within agentic ecology, although, explanation of their origins and interconnection is problematic. In fact, persistent questions about the origins of agentic modality are central to human science. Scholars ask to what degree are there stable modalities of human agency, and how do such forms and functions originate, interact, and adapt? These puzzles have been deep and widespread, especially since the European Enlightenment (Giddens, 2013). During this period, scholars elevated the status of autonomous, reasoning individuals, as well as democratic institutions, and then worked to integrate these modalities with traditional forms of family and community. This clearly contrasted the premodern emphasis on patriarchal order and cultural compliance.

© The Author(s) 2021
P. T. Bryant, *Augmented Humanity*, https://doi.org/10.1007/978-3-030-76445-6_3

Contemporary debates continue, regarding the origins and interactions of individuals, groups, and collectives. Competing answers have major implications. For example, if collective forms and functions are foundational modalities, rather than individual persons or relational groups, then collective origins take precedence. Individuals and groups will inherit many of their core characteristics from membership of cultural and social collectives. In contrast, if individual persons and their close relationships are the primitive modalities, then collectives derive from the combination or aggregation of individuals. Collectives would inherit many core characteristics from their members.

These distinctions have been major fault lines in modern thought. On the one hand, some advocate bottom-up explanations, thereby invoking methodological individualism, in which persons assemble, aggregate, or contract, into collective agentic modalities. Within theories of this kind, interpersonal comparison and negotiated consensus are frequent concerns, because they mediate a liberal approach to aggregation and combination (e.g., Arrow, 1997; Locke, 1967). On the other hand, there are those who advocate top-down explanations, thus invoking methodological collectivism, in which individuals inherit and instantiate features of the collective (e.g., Marx, 1867). Intercommunal comparison and managed consensus are now typical concerns because they mediate a cultural process of agentic devolution. Other scholars occupy the middle ground, focusing on the dynamics of relational groups, using either a sociological lens to explain how groups join into larger collectives (e.g., Simmel, 2011), or a social psychological lens to explain how group relationships shape individuals (e.g., Lewin, 1947). In almost all approaches, modern scholars accept a major role for collectives, and then debate their interaction with individuals. As March and Simon (1993, p. 13) explain, "organization members are social persons, whose knowledge, beliefs, preferences, loyalties, are all products of the social environment in which they grew up, and the environments in which they now live and work."

Agentic modalities can therefore be defined in terms of their layers of form and functional mechanisms. Notably, the hyperparameters of agentic metamodels define the same characteristics. Hence, there will be hyperparameters which specify the modalities within a metamodel of agency, including modal layers and their mechanisms of interaction, for example, in hierarchies or networks. Moreover, hyperparameters can be immediately visible, or hidden and require discovery (Feurer & Hutter, 2019). From the "persons in context" perspective, there are both visible and hidden layers and mechanisms. Much is known, but much remains to be uncovered (Cervone, 2005). Variation is contingent on context and individual difference, and perhaps the unconscious. Sigmund Freud certainly thought so, as do many of his postmodern inheritors (Tauber, 2013). In competing theories, more is visible. Persons are conceived in terms of stable, observable traits and states. From this perspective, there are fewer hidden layers and mechanisms, and less inherent variance (e.g., McCrae & Costa, 1997). Agentic modality is more visible and predictable.

Comparable distinctions apply regarding the hyperparameters of collective modality. Some theories emphasize observable structures, routines, and norms of collectivity, with few hidden layers and mechanisms. In new institutional theory, for example, organizations exemplify the observable forms and functions of institutional fields. Isomorphism, homophily, and imprinting are then predictable, because they reflect hyperparametric transparency and stability (Scott, 2014). However, in other theories, collective modality is less transparent. There are hidden layers and mechanisms which need to be uncovered, explained, and sometimes reformed (e.g., Habermas, 1991). Thinking this way, Friedrich Engels sought to expose the "false consciousness" of capitalism (Augoustinos, 1999). Intermediate processes are possible as well, in which collective layers develop through shared action and sense-making, as iterative cycles of emergence or construction (Giddens, 1984; Weick et al., 2005). In summary, each type of agentic modality entails a debate about the hyperparameters for its fundamental layers, categories, and mechanisms. All theories of agency engage with these debates, in one way or the other.

3.1 Mediators of Agentic Modality

Whether explicitly or implicitly, therefore, theories of human agency assume patterns of modal form and function. Reflecting the problematics of modernity, most offer an explanation for the relationship between individuals and collectives. Many posit a major role for procedural action in this regard, especially individual habit and collective routine. As William James (1890, p. 3) remarked, people can be described as "bundles of habits," implying that habit mediates personality. Leading contemporary psychologists agree (Wood & Rünger, 2016). Similarly, scholars view procedural routine as a key mediator of social collectives (Cohen, 2006; Salvato & Rerup, 2011). Indeed, at individual, group, and collective levels of modality, procedural patterns of action support the continuity of identity and organization (Albert et al., 2000). However, the origins of habit and routine remain problematic. At heart, the problem is one of mediated modality, as scholars debate the relationship between different layers of agency and their mechanisms of interaction (Latour, 2005). Many ask, does collective routine evolve bottom-up, from the aggregation of individual habit; or does procedural action originate at the collective level, and individual habit is then reflective of routine? Similarly, are habit and routine fixed in memory, as models or templates of action, and performances then instantiate the encoded procedure; or do habit and routine continually emerge as expressions of situated practice and performance (Pentland et al., 2012)?

In fact, the contextual dynamics of human psychology offers a way forward. To begin with, assume that a social ecology is relatively stable and endowed, sufficient to support patterns of recurrent action. As agents then interact, some share common goals and patterns of action. Over time, these patterns may become automatic among groups. In effect, the agents experience the same habituation process (Winter, 2013; Wood & Rünger, 2016). Each member of the group encodes the same triggers, procedures, and expectations of action. Moreover, each agent will encode similar social psychological processes, in the performance of action. They rely heavily on collective mind and memory, sensing the same signals from each other and the environment (see Cohen et al., 2014). Moreover, the process will not trigger significant individual differences. This is possible, because we assume that individual personality is inherently open and adaptive, and allows for the upregulation and downregulation of psychological processes (Nafcha et al., 2016). In the case of routine, many personal motivations, goals, and commitments are downregulated and effectively latent. Only a limited subset of common, psychosocial processes is upregulated and active. This subset of active, upregulated processes will often include shared encodings, beliefs, goals, and competencies, while most individual differences of these kinds are downregulated (Silver et al., 2020).

This distinction is important and worth restating. In procedural patterns of action, many individual differences, such as personal values, goals, motivations, and commitments, are downregulated and latent. Whereas, shared characteristics, such as common encodings, beliefs, and competencies, are upregulated and active. In this way, shared patterns of action emerge, which are stored in individual and collective memory, and which invoke equivalent, habitual responses among groups of people, but without activating significant individual differences. As Mischel and Shoda (1998) explain, this is how cultural norms evolve, as common, recurrent psychological processes. Hence, the formation of habit and routine is neither simply bottom-up nor top-down. Rather, it is a process of related agents downregulating their individuality, while upregulating common features of sociality. Habit and routine thus coevolve, within individual and collective modalities, respectively.

Furthermore, given the downregulation of many individual differences in routine, individual persons will be less sensitive to outcome variance in routine performance, compared to more effortful, deliberate action. They are not consciously monitoring precise expectations or aspirations. Indeed, the purpose of much habit and routine is to maintain procedural control, rather than to achieve specific goals or engage in intentional action (Cohen, 2006). Although, that said, routine and habit do adapt, in response to significant contextual change, or a major shift in beliefs or goals, and more frequently, when performance fails to achieve adequate levels of control (Feldman & Pentland, 2003; Wood et al., 2005). In these situations, individual aspirations, goals, and expectations upregulate and drive adaptation. This happens naturally, when human agents—whether individual, group, or collective—are viewed as complex, open, and adaptive systems, fully situated in context.

Issues of Combination and Choice

A major consequence of this analysis is that no mechanisms of bottom-up aggregation or top-down devolution are required to explain procedural action and collective modality. Regarding collective routine, particularly, there is no need to aggregate personal motivations, values, goals, and preferences, which is what most aggregation models seek to do (see Barney & Felin, 2013). Only a common subsystem of psychosocial functioning is upregulated, and most individual differences are down-regulated. And as stated above, this naturally occurs when individuals are conceived as complex, open, adaptive systems. Different psychological subsystems may activate or not, combine or recombine, depending on the context and stimuli. At the same time, routine action is mediated by common, social-psychological mechanisms, such as social identity, collective memory, and docility. It is via these mechanisms, that collective routine emerges as a mediated pattern of action (Winter, 2013). In fact, all types of modality could activate the same pattern of action. What distinguishes them as individual habit or collective routine, is the down-regulation and upregulation of different psychosocial processes.

It is important to acknowledge, however, that not all personalities or collectives are highly organized, and not all action is habitual or routine. Even if habit serves as a scaffold for personality, and routine serves as a scaffold of collectivity, non-procedural action regularly occurs, especially when novel, complex problems arise, and agents must be creative and innovative, or when important values and interests are at stake. Automatic, procedural routine does not suffice. In these situations, individual differences often upregulate and are salient again (Madjar et al., 2011). Agents must actively seek solutions about how to think and act. To illustrate, assume that members of a collective have strong personal preferences and expectations regarding newly offered benefits, such as access to health care and education. Personal goals and preferences are likely to upregulate in this situation. Individuals will form strong personal preferences, and the collective must negotiate how to allocate benefits among its members. This will entail an effortful process of collective choice, whereby members seek to communicate, compare, and combine their diverse preferences. More often than not, any solution will require truces and trade-offs (Cyert & March, 1992). An effortful method of collective aggregation is now required, and dilemmas of interpersonal comparison and combination quickly emerge. Ultimately, however, if this process succeeds, most members will be content, their personal differences will downregulate once again, and the outcome becomes routine. Mechanisms of routinization thereby mediate social order and organization.

In fact, this type of problem is central to social choice theory, welfare economics, and behavioral theories of organization (Arrow et al., 2010). In these fields, theories highlight the aggregation of choice, in the face of individual heterogeneity and opacity. Often, previously agreed procedures—such as voting and decision routines—allow members to reach consensus and make collective choices. Such methods enable the incomplete, but acceptable aggregation of preferences, despite contrasting interests and commitments. Scholars then debate which routine procedures should be encoded, and why (Buchanan, 2014). In practical domains, this leads to political debates about the appropriate means of collective decision-making. But importantly, most theories of this kind assume that collective modalities already exist, typically as communities and institutions.

Furthermore, once made, collective choice often becomes routine and no longer requires debate or consensus building. Indeed, as noted earlier, many natural and artificial ecologies are relatively stable and munificent over time. Communities also become accustomed to the order of things, and people value the benefits which institutional order bestows. In these contexts, many people are docile, content with procedural controls, and seek no more. Collective choice is routine, not politicized, and can be accepted with the commons (Ostrom, 1990). As a practical matter, therefore, many situations are untouched by the technical impossibility of optimal aggregation (see Arrow, 1997). Collective life proceeds fairly and effectively, without the need to debate or vote, which is good news for social cohesion and civility.

3.2   Impact of Digitalization

As preceding sections explain, artificial and human agents share numerous fundamental characteristics. Both are intelligent, goal-directed types of agent, and can be understood as complex, open, adaptive systems. Both also occur in similar patterns, as individuals, in hierarchies and networks. These similarities mean that human and artificial agents are well suited to collaborating as augmented agents. Furthermore, just like humans, artificial agents are supervised in different ways, some more plastic and self-generative. In fact, in unsupervised forms of artificial intelligence and machine learning, modality is hidden until it emerges through processing (Shwartz-Ziv & Tishby, 2017). Some artificial systems are therefore fully emergent, using highly compositive methods (e.g., Wu et al., 2010). This already happens in virtual domains (Aydin & Perdahci, 2019; Cordeiro et al., 2016). The same will be true of digitally augmented agents. We can expect to see self-generative metamodeling more widely.

However, as the complexity of data and processing increases, so do the time and resources required. Computer scientists therefore develop techniques to reduce the processing load. One major technique is the compression of modalities, that is, reducing the distinction between layers of form and function, meaning they are easier to connect and transform (Wan et al., 2017). This entails the definition of functions, categories, and system boundaries to maximize integration and the ease of interaction. Similar techniques of modal compression and modularization are also applied in organizational settings, especially those which rely heavily on digital platforms and networks (Frenken, 2006). However, these techniques entail costs. The compression of modality often increases hidden complexity, and it then takes more effort to identify and process layers and levels. In computer science, techniques have been developed to manage these challenges, including sparse sampling and partial completion (Wang et al., 2018), plus hyperparameter pruning and tuning (Tung & Mori, 2020). The goal is to generate compressed, well-fitting metamodels, while also reducing the processing load (Choudhary et al., 2020). Resulting processes are more efficient, because they require less data and fewer steps to complete.

Persistent Limitations

By contrast, human beings are limited and constrained in this regard. Their modalities are relatively layered, distinct, and slow to adapt. Indeed, human modalities tend to be stable over time. Apart from anything else, physiological and neurological evolution are relatively glacial, and will probably remain so, at least for the foreseeable future. It takes time for human beings to learn and adapt. Personalities and relationships also tend toward stability, and for good reasons. They anchor the self and group in community. Social and cultural adaptation are sluggish too. Collective norms, organizations, and institutions, all evolve relatively slowly, often requiring generational cycles. Therefore, human sluggishness and path dependence are likely to persist. Human modalities will be relatively layered and stable, compared to artificial agents.

In fact, some argue that moderate human sluggishness and path dependence are inherent and desirable in many contexts (Sen, 2018). These characteristics support the continuity of identity and meaning over time, for personalities, organizations, and cultures. They also elicit prosociality, because if human functioning is generally sluggish and incomplete, people must cooperate with each other to achieve shared goals. They cannot do so alone. Similarly, moderate intersubjective opacity often encourages trust and civility. When others are partially unknowable, people need to trust each other (Simon, 1990). Whereas the absence of such limits (actual or perceived) can lead to the over-activation of individual or group differences. And if people feel separately empowered and independent of others, then antisocial outcomes become more likely, including intolerance and oppression. In these situations, emboldened autonomy can lead to mistrust or worse. Hence, while human limitations are sometimes frustrating, needing each other promotes prosociality and community.

Reflecting these contrasting tendencies, dilemmas arise when human and artificial agents combine in augmented modalities. Their prior dispositions are resilient. Artificial agents tend to compress modality, thereby reducing the distinctions between layers of form and function, while human modalities tend to be layered and uncompressed. When both combine, therefore, artificial components could be highly compressed and flattened, and the human components are uncompressed and layered. For example, in massive online gaming, people compete against each other in a highly individualistic or group fashion, which evidences uncompressed human modality. At the same time, they collaborate with highly compressed artificial agents and avatars which interact and combine with ease (Yates & Kaul, 2019). The virtual world is compressed and flat, while the human players are layered and distinct, as individuals and teams. A risk in this context is extreme modal divergence, where the human players experience strong reinforcement of layered organization and identity, even as their artificial partners further compress. Overall coordination and performance are likely to suffer.

Second, artificial agents are increasingly self-generative, while human agents are less capable in this regard. Hence, augmented modalities might emerge in which artificial components are highly self-generative, while human components are not. Online gaming is illustrative here too. Individual personalities are relatively stable and supervised over time, while artificial agents can be highly dynamic and self-generative (Castro et al., 2018). A major risk in these situations is extreme modal convergence by over-compression. For example, players may immerse themselves too deeply and become socially disengaged, lacking a clear sense of human association and control (Ferguson et al., 2020). In fact, studies suggest that addicted players do become less sensitive to others. In more extreme online situations, people may surrender to artificial supervision and forfeit autonomous self-regulation. Key aspects of their individual functioning are downregulated and latent.

Dilemmas of Agentic Modality

Novel dilemmas therefore arise for augmented modality. These dilemmas derive from different human and artificial tendencies. On the one hand, augmented modalities could be extremely divergent, by combining static human layering with dynamic artificial compression. The topography of such modality would be equivalent to a heterogeneous landscape, covered with irregular peaks and plains. Not an easy terrain to navigate, in terms of processing (Baumann et al., 2019). In such cases, metamodels would be underfitting. That is, they would admit excessive noise and variance, and thus fail accurately to distinguish potential patterns of augmented agency (Goodfellow et al., 2016). But on the other hand, augmented modalities could be extremely convergent, by allowing artificial compression to suppress human layering. This topography would be equivalent to a smooth landscape, arguably, too easy to navigate, because metamodels would be overfitting. That is, they would omit too much noise and variance, and thus fail accurately to capture variant patterns of augmented agency. Or vice versa, augmented modalities could be extremely convergent, by allowing human layering to overwhelm and dominate modality. Now the topography would be a predictable landscape which lacks variety.

Furthermore, these effects suggest poor supervision of the entrogenous mediators discussed in the preceding chapter. Recall there are three such mediators: intelligent sensory perception, performative action generation, and contextual learning, which are critical for augmented modality. However, owing to their inherent dynamism and complexity, these mediators are difficult to supervise. They exploit rapid, intra-cyclical feedforward mechanisms, which typically elude human monitoring. They cycle quickly with high precision, and are largely inaccessible to consciousness. It is therefore difficult to involve human agents in the supervision of entrogenous mediation. Augmented modalities will easily drift toward divergence or convergence.

Ambimodality

To conceptualize this novel feature of digitally augmented modality, I import another term, "ambimodality." It comes from chemistry and refers to single processes which result in different outcome states (Yang et al., 2018). Notably, the term incorporates the prefix "ambi" once again, meaning "both." With respect to augmented agency, ambimodality refers to single processes which lead to different modal outcomes, and more specifically, processes which result in dynamic artificial compression, plus stable human layering. A system is therefore highly ambimodal when it combines both extremely compressed and uncompressed form and function. Alternatively, lowly ambimodal agents will be highly convergent, either fully compressed and dynamic in artificial terms, or fully layered and stable in human terms.

Consider the following examples. Many contemporary organizations are pursuing digital transformation. In doing so, they introduce highly compressed artificial intelligence and machine learning across the organization. However, their human employees remain uncompressed individuals and groups, layered and hierarchical. The overall result is highly ambimodal, making the organization difficult to integrate and coordinate. People and artificial agents often struggle against each other, as humans try to maintain their social identities and commitments, in an increasingly flat and fluid, digitalized environment (Kellogg et al., 2020; Lanzolla et al., 2020). Ironically, members of the organization may be increasingly connected but feel less united. Alternatively, other organizations are becoming fully virtual and digitalized, and human actors are peripheral, perhaps contract "gig" workers. The system is highly compressed and lowly ambimodal, making the organization easier to integrate and control. However, human identities and commitments are largely expunged. In fact, studies already report these effects, albeit without labeling them as ambimodal (e.g., Kronblad, 2020).

At the same time, it must be noted that ambimodal systems are not inherently dysfunctional. Human modality, whether digitalized or not, is a consistent blending of contrasts, combining stability and change, the self and the other, the one and the many (Higgins, 2006). Indeed, moderate levels of ambimodality can be advantageous in volatile, uncertain contexts. This is because, when environments are unpredictable, variable modalities enable a wider range of potential forms and functions, thereby enhancing adaptive fitness. In this respect, moderately ambimodal agents can be more robust and adaptive (Orton & Weick, 1990). In contrast, fully non-ambimodal agents generate far fewer potentials. These systems are uniformly structured and integrated. Sometimes this is beneficial, for example, in stable, technical environments. But otherwise, non-ambimodal systems tend to be inflexible and fragile. This type of risk arises in tightly bound groups (Vespignani, 2010) and in the "iron cage" of bureaucratic institutions (Weber, 2002). A major task for augmented supervision, therefore, is to maximize ambimodal fit by combining appropriate levels of modal compression and layering.

3.3   Patterns of Ambimodality

Based on the foregoing discussion, this section summarizes and illustrates the main features of digitally augmented ambimodality, and especially systems which combine extreme forms of artificial compression and/or human layering. To begin with, it is important to acknowledge that digital augmentation offers many potential benefits, for individuals, groups, and collectives. Augmented agents will possess unprecedented capabilities to compose and recompose new patterns of agency and action. If well supervised, ambimodality therefore increases agentic potentiality. In many task domains, significant benefits are already apparent. However, at the same time, it poses new risks. When human and artificial agents combine, their different characteristics can skew augmented modality. On the one hand, augmented agents could be overly divergent, by combining compressed artificial forms and functions, with more layered human forms and functions. On the other hand, agents could be overly convergent, fully dominated by artificial compression, or by human layering. In other words, there are risks of inappropriate, high or low ambimodality. Augmented agents of this kind will be less coherent and potentially dysfunctional. Recall the examples given above, of organizations which undergo digital transformation and either alienate or expel people in the process.

Low Ambimodality

In some augmented agents, there will be low ambimodality. The resulting system will be highly integrated and convergent. In fact, this type of augmented agent is like a closely knit group, but the relationships are internal, between human and artificial collaborators. Figure 3.1 illustrates the inner workings of such a system, assuming full digitalization and high modal compression. The figure builds on the generative metamodel of augmented agency, shown in Fig. 2.3. Shaded circles indicate digitalized processes, and unshaded circles are fully human. Adopting this approach, Fig. 3.1 shows two human agents A_3 and B_3, in the upper and lower portions of the figure respectively, each with three major phases of processing: input stimuli trigger sensory perception (SI and SP); followed by cognitive-affective processing, which leads to action generation (CA and AG); and then behavioral-performative outputs, which stimulate evaluation of performance (BP and EP), conditional on sensitivity to variance. Evaluation may subsequently trigger feedback encoding (FB), while feedforward encoding occurs intra-cyclically (FF). Both agents, A_3 and B_3, also combine in the

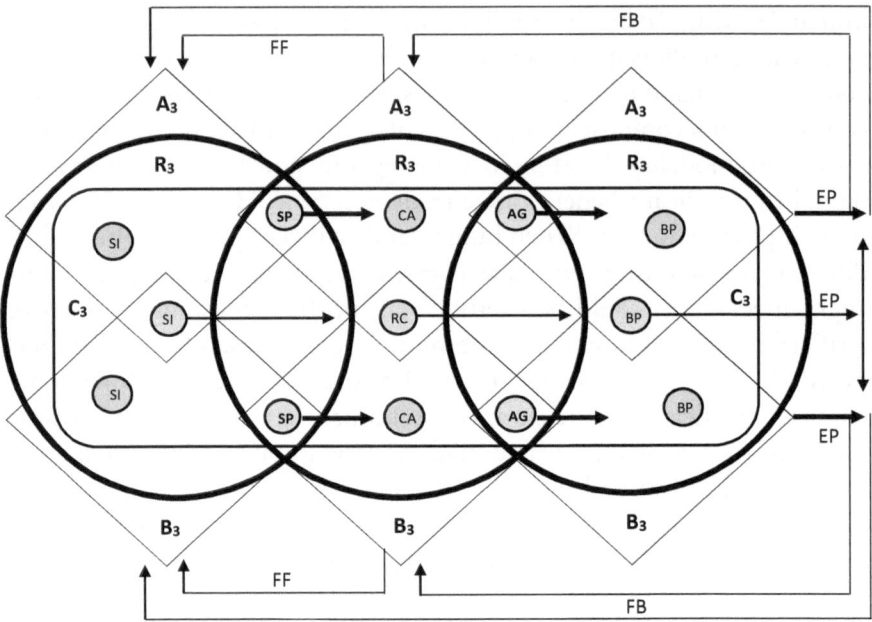

Fig. 3.1 Low agentic ambimodality

relational group R_3, which is shown by three larger, overlapping circles. Relations between phases are mediated by digitalized entrogenous mechanisms: intelligent sensory perception (SP), performative action generation (AG), and contextual learning (from FB and FF). Finally, the agents also form a collective form C_3, which spans the center of the figure.

Note that all the small circles in Fig. 3.1 are shaded. Hence, digitalized processes dominate in this scenario, and purely human processes are downregulated and latent. Human modalities are therefore compressed, shown by the lighter boundaries for human agents A_3 and B_3. Human forms and functions are less distinct. Also recall that lowly ambimodal agents are like closely knit collaborative groups. This feature is shown by the heavy boundaries for the relational group R_3 which encompasses all the digitalized processes depicted by shaded circles. Moreover, the main phases of the relationship are mediated by entrogenous mechanisms, indicated by the intersection of the large diamond shapes. In summary, Fig. 3.1 illustrates a lowly ambimodal augmented group which is highly digitalized and compressed overall.

As noted earlier, this scenario poses significant downside risks. Particularly, agentic modalities could overcompress. The downregulation of purely human functioning could go too far. Digitalized routine would overwhelm human relating and communication. Individual distinctions are effectively dissolved. If this occurs, important features of being human may be lost, or at least suppressed in this group, including the sense of autonomous agency and identity, autobiographical narratives, as well as enduring personal commitments. This type of augmented group is therefore potentially dysfunctional because many human needs and interests will be squashed by the convergent, overcompression of modality. Low ambimodality therefore presents a major challenge for the supervision of augmented agency: how to combine human layering with artificial compression, in ways which exploit and enhance the value of both while maximizing metamodel fit?

High Ambimodality

Other augmented modalities are highly ambimodal. In these scenarios, human and artificial modalities are markedly different, in terms of their compression and dynamism. Human modalities could be hierarchical and layered, while artificial modalities are compressed and flat. Forms and functions are highly distinct and divergent. Now augmented agents are like very heterogeneous groups or families, in which members are closely related but often disagree and fail to cooperate. Figure 3.2 illustrates the inner workings of this kind of system. Once again, there are two human agents labeled A_4 and B_4, each with the same three major components: input stimuli which trigger sensory perception (SI and SP); cognitive-affective processing which leads to action generation (CA and AG); and behavioral-performative outputs, which stimulate evaluation of performance (BP and EP); which may subsequently trigger feedback encoding (FB), and feedforward encoding occurs (FF). Both agents, A_4 and B_4 also combine in the relational group R_4, which is shown by the three large oval shapes. The same entrogenous mediators are central once again, indicated by the intersection of the large

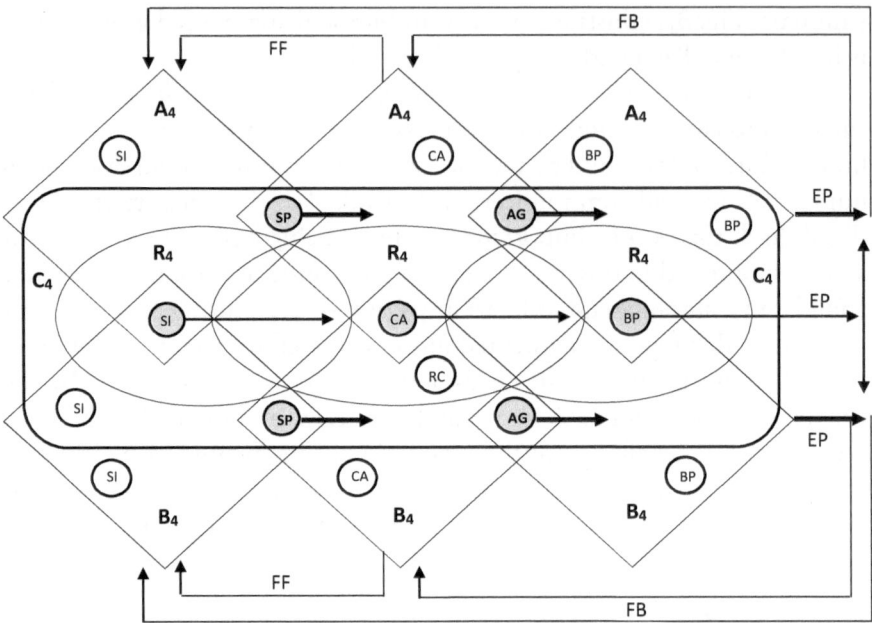

Fig. 3.2 High agentic ambimodality

diamond shapes. The agents combine in collective form C_4 which spans the center of the figure.

Digitalized processes are shaded, as before, and human are unshaded. In contrast to Fig. 3.1, however, digitalized processes do not dominate in Fig. 3.2. Human modalities are more distinct and significant. Human differences are upregulated and active. Hence, there are more unshaded circles, showing human processes, compared to the system in Fig. 3.1. Granted, the two individuals, A_4 and B_4 collaborate within relational group R_4 and collective C_4. However, individuals and groups retain greater modal distinction, compared to the system in Fig. 3.1. But in consequence, new risks appear. Human components may be highly layered, while artificial partners are highly compressed, requiring extra processing to integrate and coordinate them. At the same time, artificial agents will be highly compressed and require little effort to integrate across layers. Therefore, the combined system will exhibit different forms and functions, between human and artificial components. Overall supervision is divergent and potentially dysfunctional. In fact, as noted above, many contemporary organizations report this type of problem. They are digitally transforming many processes and systems, but their human members remain layered and cannot easily adapt (Lanzolla et al., 2020). Organizational integration and coordination are increasingly difficult to achieve. Individual differences are active, routines are fragile, and the system is harder to control. Once again, important features of being human are at risk, but now for different reasons. The persistent layering of human modality could squander the potential benefits of digital augmentation by reinforcing limiting priors. Augmentation results in ambimodal misfit and dysfunction.

3.4 Wider Implications

Throughout the modern period, scholars have assumed stable agentic forms and functions, and especially individual, group, and collective modalities. There are obvious biological and ecological reasons for doing so. Individuals, familial groups, and populations are the key organizing modalities of mammalian life (Mayr, 2002). Many theories of economics, politics, and institutions also focus on these modal distinctions, often drawing from psychology and sociology to do so. In most of these disciplines, scholars continue to debate how collectivities relate to groups and individuals. Questions remain about bottom-up versus top-down processes, and hence between methodological individualism versus collectivism, although a growing number inhabit the middle ground, theorizing about the coevolution of agentic modalities, often highlighting the role of groups and networks (e.g., Giddens, 1984; Latour, 2005).

Framing all these efforts is the modern, post-Enlightenment elevation of autonomous, rational agency. Reasoning persons took center stage, freed from the premodern strictures of superstition and autocratic order. Against this historical backdrop, the central thesis of this chapter is that mass digitalization is transforming agentic modality yet again. By exploiting digitally augmented capabilities, humanity will compose more variable forms of agentic expression and organization. Augmented modalities will be increasingly compositive and self-generative. It will also be possible to compare, contrast, and adapt modalities in a precise, dynamic fashion (Cavaliere et al., 2019). Apart from anything else, these developments challenge deeply held assumptions about the inherent opacity of reasons, preferences, and commitments (Sen, 1985). Thanks to digitalization, modality will be more transparent and composable, thereby mitigating the risk of agentic opacity for social organization.

However, as earlier sections of this chapter explain, if augmented agency is poorly supervised, modality could skew, either toward extremely convergent low ambimodality, making agents too homogeneous and lacking diversity, or toward extremely divergent high ambimodality, and agents would be too heterogeneous and lack coordination. In either scenario, digital augmentation impacts negatively on modality and degrades the efficacy of persons, groups, and collectives. Hence, the problematics of agentic modality expand from modern concerns about reductive individualism versus holistic collectivism, to include (a) concerns about artificial overcompression, combined with human overexpansion, or (b) the potential suppression of modal diversity and plasticity, and (c) the implications of these distortions for human identity, efficacy, and coherence.

Agentic Ambimodality

Among the top priorities for future research, therefore, is digitally augmented, agentic ambimodality. Recall the definition again. Ambimodality refers to single processes which result in different outcome states. With respect to augmented agency, it refers to the combination of dynamic artificial compression of form and function, with stable human layering and distinction, although, as previously noted, ambimodality is not fundamentally new, even if known by other names. But the property has not been explicitly conceptualized before, probably because its effects have been largely stable and moderate. Indeed, as noted earlier, moderate levels of ambimodality can be advantageous. For example, in highly volatile contexts and uncertain task domains, diverse modalities produce a wider range of agentic potentialities, which enhances adaptive fitness. Likewise, moderate ambimodality strengthens the resilience of personalities

(Cervone, 2005) and institutions (Kirman & Sethi, 2016), and most agents benefit from an optimal level of distinctiveness (Leonardelli et al., 2010). If anything, modern scholars explore how to encourage moderate ambimodality by developing loosely coupled, modular systems (Westerman et al., 2006).

Fresh challenges now arise because digital augmentation greatly amplifies these effects. Ambimodal extremes are more likely, as well as the dynamic composition of alternative agentic forms and functions. The full range of options was earlier shown in Fig. 2.6, which shows alternative combinations of human and artificial supervision in augmented agency. A major task, therefore, is the specification of hyperparameters for modal compression and layering, the goal being to determine the appropriate level of ambimodality in any context, and thereby to maximize metamodel fit. Otherwise, agents' inherent tendencies could lead to inappropriate extremes. These should be key topics of future research. Scholars can look to computer science for guidance, where similar topics are already major foci of research (Sangiovanni-Vincentelli et al., 2009). Management scholars are exploring these topics also, in the digital transformation of organizations (e.g., Lanzolla et al., 2020; Ransbotham et al., 2020). Some research how to embed values and commitments into the supervision of digital augmentation, for example, by clearly articulating the human purpose of systems design.

Problems of Aggregation

In numerous fields, theories posit routine as a key mediator of group and collective modalities. But questions remain about the origin and functioning of routine: does it emerge via bottom-up aggregation of habit, or does routine develop holistically and then devolve top-down, or perhaps both processes occur? These are central questions for behavioral and evolutionary theories of organizations and markets (Nelson & Winter, 1982; Walsh et al., 2006). Furthermore, many scholars in these fields argue that individuals' cognitive and empathic limitations—especially bounded rationality and intersubjective opacity—aggregate to collective limitations, compromises, and constraints. And hence, just like individuals, collectives employ procedural routine in decision-making, problem-solving, and the reading of group mind (Cyert & March, 1992). But exactly how aggregation occurs in these situations also remains a contentious puzzle (see Barney & Felin, 2013; Winter, 2013). Similar questions persist in other fields. For example, in microeconomics, scholars investigate the limits of interpersonal comparison and aggregation in collective choice (Sen, 1997). In legal theory and ethics, scholars analyze how empathic limitation shapes the organization and aggregation of commitments in contractual consensus (Sen, 2009). However, aggregation is typically imputed and not yet adequately explained.

This chapter proposes a solution, by viewing human agents as complex, open, and adaptive systems, which respond to variable contexts. From this perspective, humans naturally experience the downregulation of individual differences, in the recurrent, predictable pursuit of common goals. In parallel, they experience the upregulation of collective characteristics including social norms and control procedures. In this way, it is possible to explain the origin and functioning of individual habit and collective routine, without aggregating full personalities, personal preferences, beliefs, goals, and motivations. A common subset of mediating mechanisms does most of the work (Brinol & DeMarree, 2012). And to repeat, no special process of bottom-up aggregation or top-down devolution is required. Rather, many individual differences are downregulated and latent, while common characteristics are upregulated and active. Thus, habit and routine coevolve in procedural action.

These processes warrant deeper investigation, partly because habit and routine are prime targets for digital augmentation, but also because digital augmentation implies more dynamic processes of habit and routine (Bandura, 2007; Davis, 2015). Procedures will need to adapt and recompose, in a dynamic fashion, and adjust levels of modal compression and layering. The variable upregulation and downregulation of cognitive-affective processes will be a key to these dynamics. In these respects, habituation and routinization will require more deliberate supervision. Recent investigations into the adaptation of habit and routine offer relevant insight (Winter et al., 2012). Part of the solution will lie in identifying and managing the core components of any procedural action, and then upregulating or downregulating other factors, depending on the situation and context, to maximize metamodel fit. Digitally augmented processes will undoubtedly assist (see Murray et al., 2020). However, many questions remain unanswered.

Implications for Institutions

This analysis of routine has additional implications for social and economic institutions. For example, markets and businesses are supported by routines of production, consumption, and transaction; political institutions by routines of representation, deliberation, and decision-making; and legal institutions rely on routines of examination, judgment, and sanction. However, as this chapter explains, collective augmented agents could skew toward extreme divergence or convergence. If artificial and human components overly diverge, collectives will be internally conflicted and lack coherence. Whereas if they overly converge, they could be overdetermined by artificial agents, or dominated by inflexible human hierarchy and priors. In the meantime, social networks and virtual power are growing rapidly, but governance and trust are lagging. We see these effects already, for example, where digitalization is destabilizing the administration of politics and justice (Hasselberger, 2019; Zuboff, 2019).

In a highly augmented world, therefore, historic sources of collective coherence and consistency—such as negotiated truces, voting procedures, and routine docility—may be less effective, at least in the digitally augmented world. More will be known, transparent, and communicable, reducing the need for truces, voting, and docility. Entrogenous mediators will play a critical role here. New forms of intelligent sensory perception, performative action generation, and contextual learning, will mediate greater transparency and dynamism. Augmented agents will compose and recompose by design, rather than by imitation and other traditional means. In this respect, they will be generative and near composability, not only adaptive and near decomposability (see Simon, 1996). This contrasts prior assumptions that collective agency and choice emerge gradually, often through iterative processes of incomplete comparison and negotiation.

Viewed positively, these changes will support more agile organizations and institutions. On the downside, however, augmented collectives could over-compress and squash valued features of human experience. Alternatively, human and artificial agents might diverge and conflict, even as they collaborate more closely. In contrast, for most of human history, agentic modalities have been viewed as layered, stable forms. During premodernity, the dominant layers were communal and patriarchal, whereas, in the modern period, the most important modal layers are individual persons and social collectives. Digital augmentation problematizes these assumptions. Old stabilities and constraints are relaxing. Newer, compositive methods are now feasible, leveraging highly digitalized capabilities and networks. At the same time, fixed modal layers are giving way to more hybrid, self-generative forms. The universe of agentic modality is becoming more pluralistic and this trend is likely to accelerate. It offers genuine promise but also brings new risks. Effective supervision will be critical.

References

Albert, S., Ashforth, B. E., & Dutton, J. E. (2000). Organizational identity and identification: Charting new waters and building new bridges. *Academy of Management Review, 25*(1), 13–17.

Arrow, K. J. (1997). The functions of social choice theory. In K. J. Arrow, A. Sen, & K. Suzumura (Eds.), *Social choice re-examined* (Vol. 1). St. Martin's Press.

Arrow, K. J., Sen, A., & Suzumura, K. (Eds.). (2010). *Handbook of social choice and welfare* (Vol. 2). Elsevier.

Augoustinos, M. (1999). Ideology, false consciousness and psychology. *Theory & Psychology, 9*(3), 295–312.

Aydin, M. N., & Perdahci, N. Z. (2019). Dynamic network analysis of online interactive platform. *Information Systems Frontiers, 21*(2), 229–240.

Bandura, A. (2006). Toward a psychology of human agency. *Perspectives on Psychological Science, 1*(2), 164–180.

Bandura, A. (2007). Reflections on an agentic theory of human behavior. *Tidsskrift-Norsk Psykologforening, 44*(8), 995.

Barney, J., & Felin, T. (2013). What are microfoundations? *The Academy of Management Perspectives, 27*(2), 138–155.

Baumann, O., Schmidt, J., & Stieglitz, N. (2019). Effective search in rugged performance landscapes: A review and outlook. *Journal of Management, 45*(1), 285–318.

Brinol, P., & DeMarree, K. G. (2012). Social metacognition: Thinking about thinking in social psychology. In P. Brinol & K. G. DeMarree (Eds.), *Social metacognition* (pp. 1–18). Psychology Press.

Buchanan, J. M. (2014). *Public finance in democratic process: Fiscal institutions and individual choice.* UNC Press Books.

Castro, O. R., Fritsche, G. M., & Pozo, A. (2018). Evaluating selection methods on hyper-heuristic multi-objective particle swarm optimization. *Journal of Heuristics*, 1–36.

Cavaliere, D., Morente-Molinera, J. A., Loia, V., Senatore, S., & Herrera-Viedma, E. (2019). Collective scenario understanding in a multi-vehicle system by consensus decision making. *IEEE Transactions on Fuzzy Systems, 28*(9), 1984–1995.

Cervone, D. (2005). Personality architecture: Within-person structures and processes. *Annual Review of Psychology, 56*, 423–452.

Choudhary, T., Mishra, V., Goswami, A., & Sarangapani, J. (2020). A comprehensive survey on model compression and acceleration. *Artificial Intelligence Review*, 1–43.

Cohen, M. D. (2006). Reading Dewey: Reflections on the study of routine. *Organization Studies, 28*(5), 773–786.

Cohen, M. D., Levinthal, D. A., & Warglien, M. (2014). Collective performance: Modeling the interaction of habit-based actions. *Industrial and Corporate Change, 23*(2), 329–360.

Cordeiro, M., Sarmento, R. P., & Gama, J. (2016). Dynamic community detection in evolving networks using locality modularity optimization. *Social Network Analysis and Mining, 6*(1), 15.

Cyert, R. M., & March, J. G. (1992). *A behavioral theory of the firm* (2nd ed.). Blackwell.

Davis, G. F. (2015). Celebrating organization theory: The after-party. *Journal of Management Studies, 52*(2), 309–319.

Feldman, M. S., & Pentland, B. T. (2003). Reconceptualizing organizational routines as a source of flexibility and change. *Administrative Science Quarterly, 48*(1), 94–118.

Ferguson, C. J., Copenhaver, A., & Markey, P. (2020). Reexamining the findings of the American Psychological Association's 2015 task force on violent media: A meta-analysis. *Perspectives on Psychological Science, 15*(6), 1423–1443.

Feurer, M., & Hutter, F. (2019). Hyperparameter optimization. In *Automated machine learning* (pp. 3–33). Springer.

Frenken, K. (2006). A fitness landscape approach to technological complexity, modularity, and vertical disintegration. *Structural Change and Economic Dynamics, 17*(3), 288–305.

Giddens, A. (1984). *The constitution of society*. University of California Press.

Giddens, A. (2013). *The consequences of modernity*. Wiley.

Goodfellow, I., Bengio, Y., & Courville, A. (2016). *Deep learning* (Vol. 1). MIT Press.

Habermas, J. (1991). *Lifeworld and system: A critique of functionalist reason*. Polity Press.

Hasselberger, W. (2019). Ethics beyond computation: Why we can't (and shouldn't) replace human moral judgment with algorithms. *Social Research: An International Quarterly, 86*(4), 977–999.

Higgins, E. T. (2006). Value from hedonic experience and engagement. *Psychological Review, 113*(3), 439–460.

James, W. (1890). *Habit*. Henry Holt and Company.

Kellogg, K. C., Valentine, M. A., & Christin, A. (2020). Algorithms at work: The new contested terrain of control. *Academy of Management Annals, 14*(1), 366–410.

Kirman, A., & Sethi, R. (2016). Disequilibrium adjustment and economic outcomes. In *Complexity and evolution: Towards a new synthesis for economics*. MIT Press.

Kronblad, C. (2020). How digitalization changes our understanding of professional service firms. *Academy of Management Discoveries, 6*(3), 436–454.

Lanzolla, G., Lorenz, A., Miron-Spektor, E., Schilling, M., Solinas, G., & Tucci, C. L. (2020). Digital transformation: What is new if anything? Emerging patterns and management research. *Academy of Management Discoveries, 6*(3), 341–350.

Latour, B. (2005). *Reassembling the social: An introduction to actor-network theory*. Oxford University Press.

Leonardelli, G. J., Pickett, C. L., & Brewer, M. B. (2010). Optimal distinctiveness theory: A framework for social identity, social cognition, and intergroup relations. *Advances in Experimental Social Psychology, 43*, 63–113.

Lewin, K. (1947). Frontiers in group dynamics: Ii. Channels of group life; social planning and action research. *Human Relations, 1*(2), 143–153.

Locke, J. (1967). *Locke: Two treatises of government*. Cambridge University Press.

Madjar, N., Greenberg, E., & Chen, Z. (2011). Factors for radical creativity, incremental creativity, and routine, noncreative performance. *Journal of Applied Psychology, 96*(4), 730–743.

March, J. G., & Simon, H. (1993). *Organizations* (2nd ed.). Blackwell.

Marx, K. (1867). *Das kapital* (B. Fowkes, Trans., 4 ed.). Capital.

Mayr, E. (2002). *What evolution is*. Weidenfeld & Nicolson.

McCrae, R. R., & Costa, P. T., Jr. (1997). Personality trait structure as a human universal. *American Psychologist, 52*(5), 509–516.

Mischel, W., & Shoda, Y. (1998). Reconciling processing dynamics and personality dispositions. *Annual Review of Psychology, 49*(1), 229–258.

Murray, A., Rhymer, J., & Sirmon, D. G. (2020). Humans and technology: Forms of conjoined agency in organizations. *Academy of Management Review* (online).

Nafcha, O., Higgins, E. T., & Eitam, B. (2016). Control feedback as the motivational force behind habitual behavior. In *Progress in brain research* (Vol. 229, pp. 49–68). Elsevier.

Nelson, R. R., & Winter, S. G. (1982). *An evolutionary theory of economic change*. Harvard University Press.

Orton, J. D., & Weick, K. E. (1990). Loosely coupled systems: A reconceptualization. *Academy of Management Review, 15*(2), 203–223.

Ostrom, E. (1990). *Governing the commons: The evolution of institutions for collective action*. Cambridge University Press.

Pentland, B. T., Feldman, M. S., Becker, M. C., & Liu, P. (2012). Dynamics of organizational routines: A generative model. *Journal of Management Studies, 49*(8), 1484–1508.

Ransbotham, S., Khodabandeh, S., Kiron, D., Candelon, F., Chu, M., & LaFountain, B. (2020). Expanding AI's impact with organizational learning. *MIT Sloan Management Review*.

Salvato, C., & Rerup, C. (2011). Beyond collective entities: Multilevel research on organizational routines and capabilities. *Journal of Management, 37*(2), 468–490.

Sangiovanni-Vincentelli, A., Shukla, S. K., Sztipanovits, J., Yang, G., & Mathaikutty, D. A. (2009). Metamodeling: An emerging representation paradigm for system-level design. *IEEE Design & Test of Computers, 26*(3), 54–69.

Scott, W. R. (2014). *Institutions and organizations: Ideas, interests, and identities* (4th ed.). Sage.

Sen, A. (1985). Goals, commitment, and identity. *Journal of Law, Economics & Organization, 1*(2), 341–355.

Sen, A. (1997). Individual preference as the basis of social choice. In K. J. Arrow, A. Sen, & K. Suzumura (Eds.), *Social choice re-examined* (Vol. 1, pp. 15–37). St. Martin's Press.

Sen, A. (2009). *The idea of justice*. Harvard University Press.

Sen, A. (2018). The importance of incompleteness. *International Journal of Economic Theory, 14*(1), 9–20.

Shwartz-Ziv, R., & Tishby, N. (2017). Opening the black box of deep neural networks via information. *arXiv preprint arXiv:1703.00810*.

Silver, C. A., Tatler, B. W., Chakravarthi, R., & Timmermans, B. (2020). Social agency as a continuum. *Psychonomic Bulletin & Review, 28*, 434–453.

Simmel, G. (2011). *Georg Simmel on individuality and social forms*. University of Chicago Press.

Simon, H. A. (1990). A mechanism for social selection and successful altruism. *Science, 250*(4988), 1665–1668.

Simon, H. A. (1996). *The sciences of the artificial* (3rd ed.). The MIT Press.

Tauber, A. I. (2013). *Requiem for the ego: Freud and the origins of postmodernism*. Stanford University Press.

Tung, F., & Mori, G. (2020). Deep neural network compression by in-parallel pruning-quantization. *IEEE Transactions on Pattern Analysis and Machine Intelligence, 42*(3), 568–579.

Vespignani, A. (2010). The fragility of interdependency. *Nature, 464*(7291), 984–985.

Walsh, J. P., Meyer, A. D., & Schoonhoven, C. B. (2006). A future for organization theory: Living in and living with changing organizations. *Organization Science, 17*(5), 657–671.

Wan, Z., He, H., & Tang, B. (2017). A generative model for sparse hyperparameter determination. *IEEE Transactions on Big Data, 4*(1), 2–10.

Wang, Y., Meng, D., & Yuan, M. (2018). Sparse recovery: From vectors to tensors. *National Science Review, 5*(5), 756–767.

Weber, M. (2002). *The protestant ethic and the spirit of capitalism: And other writings*. Penguin.

Weick, K. E., Sutcliffe, K. M., & Obstfeld, D. (2005). Organizing and the process of sensemaking. *Organization Science, 16*(4), 409–421.

Westerman, G., McFarlan, F. W., & Iansiti, M. (2006). Organization design and effectiveness over the innovation life cycle. *Organization Science, 17*(2), 230–238.

Winter, S. G. (2013). Habit, deliberation, and action: Strengthening the micro-foundations of routines and capabilities. *The Academy of Management Perspectives, 27*(2), 120–137.

Winter, S. G., Szulanski, G., Ringov, D., & Jensen, R. J. (2012). Reproducing knowledge: Inaccurate replication and failure in franchise organizations. *Organization Science, 23*(3), 672–685.

Wood, W., & Rünger, D. (2016). Psychology of habit. *Annual Review of Psychology, 67*, 289–314.

Wood, W., Tam, L., & Witt, M. G. (2005). Changing circumstances, disrupting habits. *Journal of Personality and Social Psychology, 88*(6), 918–933.

Wu, K., Zhou, X.-Z., & Guo, L. (2010). Heuristic algorithm for web services composition based on interface connective relation. *Computer Engineering and Design, 31*(1), 179–183.

Yang, Z., Dong, X., Yu, Y., Yu, P., Li, Y., Jamieson, C., & Houk, K. N. (2018). Relationships between product ratios in ambimodal pericyclic reactions and bond lengths in transition structures. *Journal of the American Chemical Society, 140*(8), 3061–3067.

Yates, R. D., & Kaul, S. K. (2019). The age of information: Real-time status updating by multiple sources. *IEEE Transactions on Information Theory, 65*(3), 1807–1827.

Zuboff, S. (2019). *The age of surveillance capitalism: The fight for a human future at the new frontier of power*. Profile Books.

4

Problem-Solving

Throughout the modern period, scientific discovery, widespread education, and industrial and economic development have encouraged commitment to rationality and humanistic values as progressive forces (Pinker, 2018). Assuming such commitments, systematic reasoning and the scientific method promise resolution of increasingly complex and consequential problems. The proper ambition of problem-solving is then to transcend the limits of ordinary capability, even if rational ideals are forever unreachable. Not surprisingly, the advocates of rationality and scientific method are impatient with constraints on problem-solving and view them as challenges to be overcome. And their impatience is reasonable, given the dramatic growth of knowledge and technology during the modern period. Capabilities have greatly expanded, and many assumed limits have receded. Digital augmentation promises radically to enhance and accelerate this trend. Problem-solving continues to advance.

Even so, human capabilities remain limited. People still need to reduce potential complexity and manage cognitive load. They often do this by simplifying problem representation and/or solution search, depending on the relative significance of each activity in any problem context. This entails a series of trade-offs between accuracy and efficiency, which entail

© The Author(s) 2021
P. T. Bryant, *Augmented Humanity*, https://doi.org/10.1007/978-3-030-76445-6_4

potential costs and risks (Brusoni et al., 2007). Most commonly, the simplification of sampling and search admits distorting biases, myopia, and noise into problem-solving (Kahneman et al., 2016; Kahneman & Tversky, 2000). Granted, in some situations, such simplification is warranted and satisfactory. For example, fast and frugal heuristics often work best in uncertain or urgent situations (Gigerenzer, 1996). And when they do, the practical challenge is not to mitigate the distortions of simplification, but to maximize its effectiveness. Either way, people naturally simplify sampling and/or search, resulting in less complex problems and solutions, respectively. They do so for a range of reasons: to maximize limited resources and capabilities; because prior commitments obviate the need for comparative processing (Sen, 2005); in order to maintain cultural norms and controls (Scott & Davis, 2007); or because heuristics are most appropriate for the problem at hand (Marengo, 2015).

Herbert Simon (1979) was among the first to expose these patterns. He argues that to solve problems with bounded or limited rationality, people simplify different aspects of problem-solving and satisfice at lower levels of aspiration, rather than fully satisfying criteria of optimality. Simon (ibid., p. 498) identifies two broad types of satisficing in problem-solving: "either by finding optimum solutions for a simplified world, or by finding satisfactory solutions for a more realistic world." On the one hand, that is, agents simplify the representation of problems, to reach optimal solutions. In this case, the processing required is more owing to the complexity of solution search. The major risks are myopic sampling and problem representation. Following common naming conventions, I call this type of problem-solving normative satisficing (see Simon, 1959). On the other hand, agents simplify solutions and address more realistic, better described problems. In this case, the processing required is more owing to the complex representation of the problem itself. Now the major risks are myopic solution search and selection. I call this type of problem-solving descriptive satisficing, again following convention.

Notably, the latter approach—accepting satisfactory solutions to more realistic problems, or what I call descriptive satisficing—is the type of problem-solving found in behavioral theories of decision-making, economics, and organizations. Stated in more formal language, it seeks no worse solutions, to the best representation of problems. Whereas, the former approach—seeking optimal solutions to simplified problems, which I call normative satisficing—is typical of classical microeconomics and formal decision-making (March, 2014). Put more formally, it seeks the best solutions, to no worse representation of problems. Hence, as Sen (1997b) explains, satisficing can be conceived as a type of formal maximizing, meaning problems and/or solutions are partially ordered, and agents accept some no worse option as good enough, assuming an aspiration level.

However, while descriptive satisficing is widely studied, normative satisficing is not. Even though Simon explained this important distinction decades ago—that classical theory also satisfices in the normative sense, by seeking optimal solutions for a simplified world—few studies investigate this phenomenon. Levinthal (2011, p. 1517) also observes this oversight, when he writes that all "but the most trivial problems require a behavioral act of representation prior to invoking a deductive, 'rational' approach." Yet despite his astute observation, with a few notable exceptions (e.g., Denrell & March, 2001; Fiedler & Juslin, 2006), most behavioral researchers focus on descriptive satisficing, that is, finding satisfactory solutions for a more realistic world (e.g., Kahneman et al., 2016; Luan et al., 2019). Granted, this is an important topic. However, as a consequence, we still await a full treatment of bounded realism, representational heuristics, and normative satisficing, especially in classical theory (Thaler, 2016). This is another large project, but I will not attempt to fill the gap here.

Historical Developments

These questions have a history worth recounting. For over two centuries, classically inspired economists have idealized Adam Smith's (1950) notion of the invisible hand to explain collective, calculative self-interest (Appiah, 2017). Equally, they idealize his characterization of *Homo economicus*, as a rational egoist bent on optimizing utility. Here are the roots of normative satisficing in microeconomics: seeking optimal, calculative solutions to simplified problems of economic utility. However, Smith (2010) also understood that rational egoism is a fictional, albeit functional ideal. In parallel, he recognized the complexity of human sentiments and commitments. From this more realistic perspective, *Homo economicus* defers to *Homo sapiens* meaning a richer conception of human agency and psychology (Thaler, 2000). Therefore, Smith also set the agenda for descriptive satisficing: accepting satisfactory solutions to well described, realistic problems, including problems of preferential and collective choice. In recent years, more scholars are embracing this broader conception of economic agency, responding to the increasing complexity and variety of choice (e.g., Bazerman & Sezer, 2016; Higgins & Scholer, 2009), although, as noted above, most research is still framed in terms of descriptive satisficing and largely overlooks the puzzles of normative satisficing, including myopic sampling and representational heuristics. Notable exceptions exist in the literature, but they are exceptional (e.g., Fiedler, 2012; Ocasio, 2012).

By contrast, the problems of both normative and descriptive satisficing are central to modern scientific method. Experimental researchers have consistently refined their methods of attention, observation, sampling, and problem representation. Indeed, the technological enhancement of attentional focus and observation are central to scientific method, along with the enhancement of data analysis and solution search. For example, in the early modern period, the telescope and microscope revolutionized observation in astronomy and biology, respectively. Using these tools and techniques, novel problems emerged which rendered prior explanations obsolete. In parallel, new mathematical and statistical methods enabled deeper analysis. Fast forward to the present, and observational tools include satellites, particle accelerators, and quantum microscopy. At the same time, computing technologies massively enhance the compilation and analysis of observational data. Using these techniques, today's scientists represent and solve increasingly novel, complex, highly specified problems. Natural science continues to transcend the limits of human capabilities and consciousness, especially in the sampling and representation of problems.

Not surprisingly, social and behavioral scientists attempt to do the same (Camerer, 2019). In these fields, however, selective sampling and experimental techniques prompt concerns about oversimplification and validity. Many caution that social and behavioral phenomena are too variable and situated, and cannot be reduced to measurable constructs and mechanisms (e.g., Beach & Connolly, 2005; Geertz, 2001). Regarding problem-solving, particularly, some argue that this activity is best explained in terms of narrative interpretation and sense-making, rather than rational expectations, preference ordering, and reasoned choice (e.g., Bruner, 2004; Smith, 2008). By implication, determinant models of problem-solving will be overly simplified and mired in assumptions of normality and stability. Others are somewhere in between. They still present formal models and methods, but embrace a broader psychology of commitments, including empathy and altruism (e.g., Ostrom, 2000; Sen, 2000). As a further example, Stiglitz et al. (2009) argue for a richer description of human needs and wants, shifting toward *Homo sapiens*, and demonstrate how these could be measured and analyzed. Their ambition is an economics of human flourishing and well-being, with public policies to match.

Nevertheless, like most, these scholars agree that something must be simplified, to develop useful theories and actionable knowledge. Debate then focuses on what to sample, simplify, and conceptualize, when and how, and with what consequences for problem-solving. As stated above, those who endorse classical theory tend to simplify problems and psychology, seeking to optimize calculative solutions, whereas behavioral approaches seek richer problem representation, and then accept approximating heuristic solutions. The debate exemplifies the modern problematic noted in earlier chapters: to what degree can and should human beings overcome their limits, to be more fully rational, empathic, and fulfilled?

Contemporary Digitalization

Digitalization now brings the advanced capabilities of empirical science and computer engineering to everyday, human problem representation and solution. For example, consider personal digital devices, such as smartphones and tablet computers. They grant individuals access to increasingly powerful and intelligent sampling, search, and computation, far beyond traditionally bounded capabilities. Using such devices, humans become collaborators in digitally augmented problem-solving. Importantly, these capabilities also reduce the need for trade-offs. Less must be simplified. Artificial agents can process the enormous amount of information required to analyze highly complex problems and choices, and at all levels of agentic modality including collectives (Chen, 2017). In augmented collaboration, therefore, humans will have the potential to behave fully as *Homo sapiens*. In fact, it becomes feasible to pursue highly discriminate problems and solutions in many ordinary contexts, not just in the laboratory (Kitsantas et al., 2019). Thanks to digital augmentation, much human problem-solving will approach scientific levels of detail, precision, and rigor, in both sampling and search.

Yet at the same time, natural human capabilities remain limited and parochial values and commitments will likely persist. Given these enduring features of human problem-solving, digital augmentation may compound rather than ameliorate behavioral dilemmas. For example, if racial and gender biases are encoded into training data and algorithmic processing, machine learning leads to even greater discrimination. Digitally augmented capability amplifies biased beliefs about gender and race (Osoba & Welser, 2017). As another illustration, consider classically inspired economics, in which problem-solving is often assessed in terms of the rational optimization of self-interested utility. Here too, digital augmentation could lead to increasingly dysfunctional problem-solving, if augmented agents simply reinforce narrow assumptions about self-interest and expectation, and overlook wider ecological, social, and behavioral factors (Camerer, 2019; Mullainathan & Obermeyer, 2017). Digitally augmented capability would thus amplify the idiosyncratic noise which often clouds decision making (Kahneman et al., 2021). It is therefore appropriate to ask, under which conditions will digital augmentation enable more effective problem-solving, rather than perpetuating the limiting myopias and models of the past; and hence, which additional procedures might help to minimize the downside risks of digital augmentation, while maximizing the upside? Moreover, these questions are urgent. Already, the speed and scale of digital innovation are transforming much problem-solving. Organizations, institutions, and citizens are struggling to keep up, trying to remain active and relevant in the supervision of these digitally augmented processes.

4.1   Metamodels of Problem-Solving

To analyze the digital augmentation of problem-solving more deeply, we first need to review dominant metamodels of problem-solving, that is, the major problem-solving choice sets. As the preceding discussion explains, modern approaches combine two main functions: sampling of various kinds, which results in problem representation, followed by solution search and selection. Both functions—problem sampling and representation, and solution search and selection—can be more, or less, specified and complex. In ideal, optimal problem-solving, each should be fully specified and result in the best possible option, although this is rarely achieved and often impossible in practical contexts. In this regard, ideal problem-solving is truly an ideal, whereas people function with limited resources and capabilities. Given these constraints, a few metamodels of problem-solving are possible.

First, as Simon (1979) explains, agents can seek optimal solutions to simplified problems, that is, the normative satisficing of classical theory. Often, such solutions are axiomatic and formalized, while problems are represented in clear, but simplified terms. Hence, from a critical behavioral perspective, "utility maximization" is a simplified representation of the problem of economic choice. And highly calculative solutions to such problems—such as rational or adaptive expectations—can only aspire for optimality because of normative satisficing. If people choose this metamodel, more processing is required, owing to the complexity of optimizing the solution. As noted earlier, the major risks of doing so are the distortions which arise from myopic sampling and simplified problem representation.

Second, agents can seek satisfactory solutions to more fully described, realistic problems, that is, descriptive satisficing. Solutions are frequently heuristic and approximate, while problems are represented in a more detailed fashion. In this type of satisficing, solutions are partially ordered, while problem representation is highly discriminated, striving for completeness. Hence, problem representation is optimized, meaning the chosen representation is the best alternative. While the chosen solution is maximal, that is, no worse than alternatives. If people choose this metamodel, more processing is required, owing to the complexity of the problem itself. Major risks arise from myopic solution search and simplified selection criteria.

Third, it is at least conceivable to seek optimal solutions to realistically represented problems, that is, ideal problem-solving which is not satisficing in either sense. However, as noted earlier, this type of problem-solving is rarely observed in human contexts, and arguably impossible, in practical terms. Nonetheless, ideal metamodels are conceivable and play significant roles in abstract thought and formal approaches (March, 2006). Both problem representations and solutions are, in principle, fully ordered, and the best options are selected. Hence, I describe this metamodel as ideal optimizing. However, if people try to apply it in practice, they typically fall short owing to high complexity and limited capabilities, which is not to say they should not try. As March and Weil (2009) argue, pursuing unreachable ideals has its place, by helping to inspire and engage agents in the face of uncertainty and resistance.

Fourth, agents seek satisfactory solutions to simplified problems, which is not satisficing either, because agents do not seek to optimize problem representation or solution. Instead, both are no worse, at best, given some aspiration levels. In fact, this is the most frequent and feasible metamodel of problem-solving in practical terms (Sen, 2005). Much of the time, people solve imprecise problems in imprecise ways, which is good enough. Hence, we can expand Simon's original analysis. As he correctly explains, there are situations in which agents rightly pursue optimal problem representation or optimal solutions, and one or the other might be attainable, or at least approachable. However, these satisficing options are an important subclass of problem-solving, not the full universe. Much practical problem-solving is not optimizing in either respect. It is fully maximizing instead, often owing to high complexity, or because optimizing is simply unwarranted. However, such problem-solving is therefore doubly myopic, in both problem sampling and representation, and solution search and selection. I label this metamodel of problem-solving as practical maximizing: choosing incompletely ordered, satisfactory solutions, to incompletely ordered, simplified problems. From this perspective, many "fast and frugal heuristics" are instances of practical maximizing (see Gigerenzer, 2000).

It is also important to note, that much practical maximizing is procedural, performed as individual habit or collective routine. The everyday world presents many ordinary problems which are appropriately solved in this way. Moreover, as the preceding chapter explains, this type of problem-solving is central to the coherence and organization of agentic modalities. In fact, many modalities come into being as systems of practical maximizing in problem-solving. Fortunately, enough problems are recurrent, easily recognized, and require little analysis to resolve. Habitual and routine problem-solving are sufficient. Not surprisingly, agentic modalities cohere around patterns of such procedures. Bundles of habits then mediate personalities, and bundles of routines mediate organizations. There are fewer processing trade-offs in both scenarios because less effort is required. Procedural, practical maximizing is efficient and sufficient. That said, failures still occur and are impactful, because habit and routine often fulfill important control functions. Maximal does not mean minimal or trivial, but rather less than optimal.

Figure 4.1 summarizes the four major metamodels of problem-solving just described. The figure's dimensions are the complexity of problems and the complexity of solutions. Both range from high to low. The figure also assumes that complexity is proportional to the degree of variation,

Complexity of Solutions

		High	Low
		1	2
Complexity of Problems	High	I - Ideal Optimizing (best problem, best solution)	D - Descriptive satisficing (best problem, no worse solution)
		3	4
	Low	N - Normative satisficing (best solution, no worse problem)	P - Practical Maximizing (no worse problem, no worse solution)

Fig. 4.1   Metamodels of problem-solving

and hence to the processing required for rank ordering. That is, the more varied the choice set, the more complex it is, and the more processing required to discriminate between options. Given these assumptions, quadrant 1 depicts the ideal optimizing metamodel or the best solution for the best representation of problems. Quadrant 2 shows descriptive satisficing, which is seeking satisfactory solutions to the best representation of problems. Next, quadrant 3 shows normative satisficing, or seeking the best solutions to simplified problems. Finally, quadrant 4 shows practical maximizing, which is seeking satisfactory solutions to simplified problems. In this final metamodel, problem representation and solution search are both no worse than the alternatives, and hence maximizing on both dimensions.

4.2   Dilemmas of Digital Augmentation

Many digital processes need to complete as quickly and accurately as possible and must avoid unnecessary processing. For example, consider the artificial agents which manage high reliability operations, monitor human safety, and mediate online transactions. They must function rapidly, with high accuracy, yet at the same time, gather and analyze massive volumes of data. Computer scientists therefore research how to maximize the efficiency of their processing. Adding to the challenge, artificial agents can easily over-sample problems, over-compute solutions, and over-complete rank ordering (e.g., Lee & Ro, 2015). Granted, overprocessing is sometimes beneficial. It can enhance robustness, by generating a richer set of options and thus slack. However, the risk is that processing becomes overly complex and less efficient. These are central issues for the design and supervision of artificial agents.

To mitigate these risks, artificial agents also simplify problem-solving. They accomplish this using algorithmic hyperheuristics and metaheuristics, defined as shortcut means of specifying metamodel hyperparameters and model parameters, respectively (Boussaid et al., 2013). Hyperheuristics are first used to compose choice sets of potential models of problem-solving and thereby to define metamodels of problem-solving, for example, composing sets of calculative or associative approaches (Burke et al.,

2013). Next, given the resulting metamodel, metaheuristics are employed to select the appropriate model for solving a particular problem (Amodeo et al., 2018). The chosen model is then applied to resolve the focal problem, for example, using specific heuristic procedures. Importantly, at each level of processing, simplifying heuristics helps to manage the complexity of processing. As Chap. 1 also explains, research in artificial intelligence focuses on optimizing such hierarchies of heuristics. They are critical for the efficiency and effectiveness of problem-solving.

Human agents do likewise, although often unconsciously and automatically. They use simple, often routine hyperheuristics and metaheuristics. When faced with a new problem, a person may unconsciously deploy an encoded hyperheuristic to specify the appropriate metamodel of problem-solving (Fiedler & Wanke, 2009). For example, the context may be familiar and uncomplicated, suggesting a simplified, heuristic approach. Next, the person will apply a metaheuristic to choose one specific model. Perhaps the focal problem reflects prior experience and can be solved using limited sampling. Fast and frugal heuristics procedures could work very well. Studies of "gut feel" in decision-making exhibit this pattern (Gigerenzer, 2008).

Digital augmentation is now transforming this domain. For example, many experts use real-time, decision support systems powered by artificial intelligence (McGrath et al., 2018). Human intuition and calculation are being digitally augmented, and it may no longer be necessary or appropriate to reply solely on human inputs. In fact, Herbert Simon (1979, p. 499) predicted this shift many years ago: "As new mathematical tools for computing optimal and satisfactory decisions are discovered, and as computers become more and more powerful, the recommendations of normative decision theory will change."

The challenge for human agents is learning how to integrate these additional sources of information and insight into problem-solving. However, given the complexity and speed of artificial processes, they are often opaque (Jenna, 2016). In fact, the inner workings of complex algorithms mirror the opacity of the human brain. It may be impossible to know exactly what artificial agents are doing, especially in real-time processing. In these respects, natural and artificial neural networks are deeply alike. Both employ extremely complex, dynamic connections, which are difficult to monitor, supervise, and predict (Fiedler, 2014; He & Xu, 2010). That said, the similarity of both agents increases the likelihood of developing effective methods of integration and supervision. Human and artificial agents are feasible collaborators in augmented agency.

Risk of Farsighted Processing

Furthermore, as collaborative capabilities become more powerful, augmented agents might err toward overly farsighted sampling and search, the opposite of myopia. They could easily over-sample the problem environment, search too extensively, and then over-compute solutions. Where farsighted in this context, is not simply a reference to spatial distance, but any sampling or search vector. To conceptualize this effect, I borrow a term from ophthalmology, "hyperopia," which means farsighted vision, the opposite of nearsighted myopia (Remeseiro et al., 2018). When hyperopia occurs in problem-solving, agents will sample in a farsighted fashion to represent problems in rich detail; or they can search for solutions in a farsighted, extensive way. In both cases, hyperopia increases the overall complexity of problem-solving (Boussaid et al., 2013). Computer scientists already research ways to avoid these risks. Better supervision is key.

In contrast, with a few notable exceptions once again (see Denrell et al., 2017; Fiedler & Juslin, 2006; Liu et al., 2017), studies of human problem-solving largely ignore hyperopic risks. This neglect is partly owing to the priorities discussed earlier, namely, that scholars traditionally focus on myopia and limited capabilities. Moreover, when myopia is the primary focus of concern, farsighted sampling and search (that is, hyperopia) may be a welcome antidote. If so, then a modest degree of hyperopia is not a problem, but a potential advantage (e.g., Csaszar & Siggelkow, 2010). Moreover, when farsighted sampling and search do occur, they are typically viewed as natural characteristics relating to perceived temporal, spatial, and social distance (Trope & Liberman, 2011). For all these reasons, hyperopia is rarely included in studies of human problem-solving, and almost never viewed as problematic. However, in a world of digitally augmented capabilities, hyperopia is likely and potentially extreme. Going too far becomes a significant risk.

That said, neither myopia nor hyperopia is inherently erroneous. Indeed, depending on the problem context, if commitments, values, and interests are well served, and if satisfactory controls are maintained, then myopic or hyperopic, sampling and/or search, can be appropriate and highly effective (e.g., Gavetti et al., 2005). This is true for human and artificial agents alike. For example, when problems are stable and recurrent, myopic sampling and search may be fully suited to the task (Cohen, 2006). This is often the case in habitual and routine problem-solving. Alternatively, when problems are complex, multidimensional, and not urgent, then hyperopic processes could be more appropriate (Bandura, 1991; Forster et al., 2004). This is often the case in technical problem-solving and replication studies.

Dilemmas of Hyperopia

Notwithstanding these exceptions, humans tend to be myopic in sampling the problem environment and searching for solutions (Fiedler & Wanke, 2009). People must be trained, therefore, to overcome their natural myopia. Many areas of education and training focus on doing this, trying to develop capabilities, training students to sample and search more widely in specific domains. Moreover, if this training is successful, lessons are deeply encoded. Yet such learning is problematic in digitalized contexts. This is because digitally augmented capabilities increasingly transcend human limitations. Consequently, hyperopic efforts may be redundant because this is what digitalized systems are good at. But people may continue striving to overcome their limits and myopias—to extend problem sampling and solution search—irrespective of the extra capabilities acquired through digital augmentation. Trained to overcome limits, they continue reaching for hyperopia, trying to be more farsighted despite the fact, that digitally augmented processes already do so. The overall result is likely to be excessive sampling and search, or extreme hyperopia. What was corrective in previously myopic contexts is now a source of hyperopic distortion.

Artificial agents face complementary challenges in this regard. Although, in contrast to humans, artificial agents are built to be hyperopic, to sample and search widely, gathering massive volumes of information, and then process inputs at great speed and precision. Hence, artificial agents also tend toward hyperopic sampling and search, but by design. For this reason, they can also go too far and be overly hyperopic, which increases complexity and reduces efficiency. When these dispositions are imported into augmented agency, artificial and human agents easily compound each other. Human agents are trained to go further, and artificial agents go further by design. Hence, computer scientists research how to prevent over-sampling and over-computation, and to limit hyperopia (Chen & Barnes, 2014). This has led to a range of technical solutions, including hyperheuristics and metaheuristics, and algorithmic constraint satisfaction (Amodeo et al., 2018; Lauriere, 1978). In fact, a recent study conceptualizes "constraint satisficing," mimicking Simon's work in behavioral theory (Jaillet et al., 2016). Problems still arise, however, when simplifying heuristics are infected by human myopias and other priors (Osoba & Welser, 2017).

Myopia with Hyperopia

For complementary reasons, therefore, both human and artificial agents are trained to supervise the upper and lower bounds of complexity in problem-solving. On the one hand, human agents are naturally myopic and trained to do more. While on the other hand, artificial agents are naturally hyperopic and trained to do less, in specific contexts. Therefore, both agents move in opposite directions and are trained to correct in opposing ways, especially in complex problem-solving. Given these divergent characteristics and strategies, if collaborative supervision is poor, they will easily undermine each other and reinforce problematic tendencies. Artificial agents could be overly hyperopic, while their corrective procedures reinforce human myopia (Balasubramanian et al., 2020). At the same time, humans could remain myopic, while their corrective procedures reinforce artificial hyperopia. In this fashion, inadequate supervision of augmented agency could lead to problem-solving, which is highly myopic in some human respects, and highly hyperopic in artificial ways. Stated otherwise, augmented agents could be extreme satisficers. Either seeking overly optimized solutions to overly simplified problems (extreme normative satisficing) or accepting overly simplified solutions to overly detailed representations of problems (extreme descriptive satisficing).

Once again, we see the effects of poorly supervised, entrogenous mediation. Ideally, augmented agents will use these dynamic capabilities to maximize metamodels of problem-solving, adjusting sampling and search to fit the problem context. As noted earlier, however, the supervision of such capabilities will be challenging, given the speed and precision of digitalized updates. Each agent will struggle to monitor the other. Distorted outcomes are likely if the supervision of entrogenous mediation is poor. First, intelligent sensory perception might simply reinforce human myopia. For example, when racial biases guide hyperopic sampling in machine learning, algorithms are quickly discriminatory (Hasselberger, 2019). Second, if the supervision of performative action generation is equally poor, it could reinforce existing procedures, for example, by escalating racially biased behaviors. And when both effects occur, human myopia and artificial hyperopia will compound to produce dysfunctional, highly discriminatory problem-solving.

To conceptualize these effects, I adopt another ophthalmic term "ambiopia" which means double vision, which is also referred to as "diplopia" (Glisson, 2019). In these conditions, the same object is perceived at different distances by each eye—one being nearsighted and myopic, the other farsighted and hyperopic—causing the agent to perceive the image with distorted, double vision (Smolarz-Dudarewicz et al., 1980). Moreover, like other novel terms in this book, "ambiopia" includes the prefix "ambi" which is Latin for "both." In the diagnosis of vision, ambiopia refers to the compounding of visual distortions. Analogously, in problem-solving, it can refer to the compounding of nearsighted myopia with farsighted hyperopia in problem representation and/or solution search.

Summary of Digitalized Problem-Solving

Based on the preceding discussion, we can summarize digitally augmented problem-solving. First, like human agents, artificial agents perform two key processes: sampling and data gathering, leading to problem representation; then searching for and selecting solutions to such problems. Humans are limited in these respects and tend to be myopic, while artificial agents are capable of increasingly complex, hyperopic sampling and search. In fact, digital augmentation imports the hyperopic methods of experimental science into ordinary problem-solving. Second, human and artificial agents both use heuristics to choose between alternative logics, models, and procedures of problem-solving (Boussaid et al., 2013). Heuristics are often layered, in a hierarchy of increasing specificity, from hyperheuristics in the specification of metamodels to metaheuristics about models and then heuristics for specific solutions. Third, human agents often encode ontological, epistemological, and normative commitments into artificial agents, which then guide sampling and search and which help to establish the threshold of sensitivity to variance, although such priors are frequently distorting, when myopia and bias are amplified by artificial means. Fourth, both types of agents need to manage the risks of myopic and hyperopic sampling and search, balancing the demands of speed, accuracy, efficiency, and appropriateness. Otherwise, human myopia and artificial hyperopia will combine to produce dysfunctional, ambiopic problem-solving.

The goal for augmented agents, therefore, is to maximize metamodel fit in any problem context. This will entail adjusting the relative myopia and hyperopia of sampling and/or search, combining both human and artificial capabilities, although, as noted, such supervision will be challenging. Instead, agents' inherent tendencies will often lead to extreme patterns, either combining persistent human myopia with unfettered artificial hyperopia (extreme divergence and high ambiopia) or allowing one agent fully to dominate the other (extreme convergence and low ambiopia). Notably, these risks are already topics of research in artificial intelligence (Amodeo et al., 2018; Burke et al., 2013). Computer scientists worry about these problems already. What is not yet adequately understood is how these processes impact problem-solving by human-artificial augmented agents.

4.3   Illustrative Metamodels

Building on the preceding discussion, the following sections illustrate two major metamodels of digitally augmented problem-solving, that is, metamodels which are highly digitalized, similarly to the generative metamodel of agency shown in Fig. 2.3. The first scenario illustrated below is a highly ambiopic metamodel of problem-solving, with very divergent levels of complexity and simplification, in problem representation and solution search. The second is a non-ambiopic metamodel with very convergent levels of complexity and simplification. While these two scenarios are not exhaustive, they highlight ambiopic risks, associated mechanisms, and their consequences.

Highly Ambiopic Metamodels

Figure 4.2 illustrates highly ambiopic metamodels of problem-solving. The vertical axis shows the level of complexity of problem representation, and the horizontal axis shows the complexity of solution search. Both range from high complexity to relative simplicity. Where high complexity implies a hyperopic, farsighted process, and low complexity (or simplification) implies a myopic, nearsighted process. In addition, the figure shows two levels of processing capability, depicted by curved lines. One is labeled L_2, and represents the processing capability of modernity, which assumes relatively moderate levels of technological assistance. The second is labeled L_3, representing the greater processing capabilities of digital augmentation, now assuming high levels of technological assistance.

As the figure shows, the greater the processing capabilities, the greater the complexity of problem representation and solution search. Capabilities and achievable complexity are positively associated. Nevertheless, even digitalized capabilities remain limited to some degree, meaning they need to be distributed. Figure 4.2 depicts this type of distribution. It also shows that combined complexities reach limiting asymptotes of complexity, for both problem representation and solution search. These upper limits are almost never reached, in practical terms. But they do play a significant role in formal modeling, and by setting the upper bounds of problem representation and solution search.

Descriptive Satisficing The next features of Fig. 4.2 to note are the segments within it. To begin with, recall that descriptive satisficing is defined as seeking satisfactory solutions to the best representation of problems (see quadrant 2 in Fig. 4.1). Now consider D_2 in Fig. 4.2 which depicts such a metamodel, assuming modern processing capabilities L_2. Problem representation is moderately complex, and the solution is simplified. Overall, therefore, D_2 is moderately divergent and ambiopic. It also suggests that solution search is anchored (and hence semi-supervised) by human myopia. This type of problem-solving is common in behavioral and informal approaches. Also note that D_2 intersects with a small, curved section of L_2. This feature illustrates a degree of possible variance in problem-solving, or in other words, the satisficing nature of such problem-solving. By contrast, the segment labeled P_2 depicts practical maximizing, given capabilities L_2. This type of problem-solving was pre-

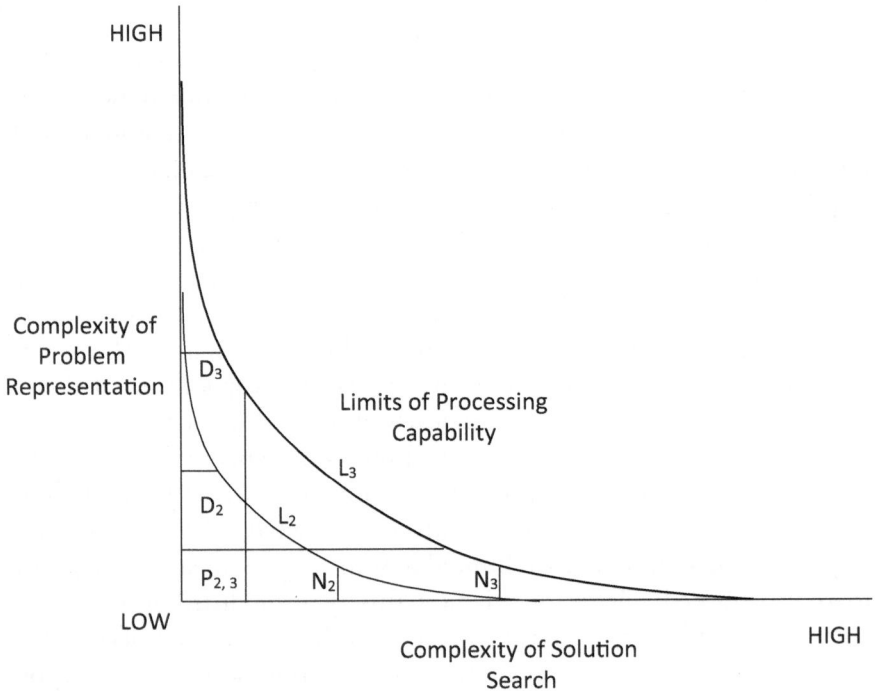

Fig. 4.2 Highly ambiopic problem-solving

viously defined as the simplified representation of problems, combined with the search for satisfactory, simpler solutions (see quadrant 4 in Fig. 4.1). Therefore, the levels of complexity are relatively low and roughly equal in P_2, meaning this type of problem-solving is non-ambiopic. Neither sampling nor search is particularly hyperopic; rather, both are relatively myopic. In fact, P_2 illustrates the actual problem-solving of most agents in a modern, behavioral world. People are not optimizing in either sense, but they rather solve simplified problems in efficient ways, often using heuristic and intuitive means. Practical problem-solving is often like this.

Next, consider the segment labeled D_3 in Fig. 4.2, which assumes stronger digitalized capabilities at level L_3. Here too, the segment denotes descriptive satisficing. However, D_3 shows greater divergence between the complexity of problem representation and simplification of solution search, and D_3 is therefore highly divergent and ambiopic overall. Agents are now digitally augmented and capable of more complex problem representation and satisfactory solution search. They sample in a farsighted, hyperopic fashion, but solution search remains anchored in the same myopias as D_2, for example, when racially biased priors are encoded into machine learning algorithms. In consequence, D_3 constitutes an extreme form of descriptive satisficing.

Finally, the segment labeled P_3 depicts practical maximizing, assuming digitally augmented capabilities L_3. That is, P_3 depicts incompletely ordered, simpler solutions, to incompletely ordered, simpler problems. As in P_2, the levels of complexity in P_3 are roughly equal, meaning this type of problem-solving is non-ambiopic, but it is highly myopic overall. In fact, P_3 almost equals P_2. This is because prior anchoring commitments have not shifted but are carried over from modernity to digitalization. This scenario illustrates the persistence of ordinary commitments and procedures. Even with digitalized capabilities, people do not seek to optimize, in either sense, but continue to rely on heuristic and intuitive means. They are persistently human, notwithstanding digital augmentation.

Normative Satisficing Now consider the other set of segments in Fig. 4.2. To begin with, recall that normative satisficing is defined as optimal solutions to simplified problems. Segment N_2 depicts such a metamodel, assuming modern processing capabilities L_2. Within N_2, the figure shows moderate divergence between the two dimensions of complexity—problem representation and solution—and therefore N_2 is moderately divergent and ambiopic overall. As noted earlier, this type of problem-solving is often axiomatic and calculative, as in classical economics: seeking optimal, calculative solutions to simplified problems of utility. Whereas the segment labeled P_2 again depicts practical maximizing, given modern capabilities L_2. It illustrates non-ambiopic, actual problem-solving in the modern world of consumption and exchange. In such a world, most people often do not optimize nor try to. Rather, they solve the ordinary problems of transactional life using habitual or routine, heuristic, and intuitive means.

Next, consider the segment labeled N_3, which depicts extreme normative satisficing, assuming stronger, digitalized capabilities L_3. In this kind of problem-solving, artificial processes enable hyperopic search, but problem representation remains anchored in the same myopias as N_2. Therefore, N_3 shows even greater divergence between the two levels of complexity, and N_3 is highly divergent and ambiopic overall. In fact, this type of distorted problem-solving is observed in semi-supervised machine learning, when hyperopic artificial intelligence amplifies human myopia and bias (Osoba & Welser, 2017), whereas the segment labeled P_3 depicts practical maximizing, given capabilities L_3. As in P_2, the levels of complexity in P_3 are roughly equal, meaning this problem-solving is non-ambiopic. In these respects, P_3 illustrates the actual problem-solving of augmented agents in the behavioral world. Consider, for example, how many people search the internet or shop online, saving favorites and encoding habits.

Descriptive and Normative Satisficing   In poorly supervised augmented agents, both types of extreme satisficing (descriptive D_3 and normative N_3) are likely and will often occur together. Persistent human priors will be myopic and artificial hyperopia will be largely unchecked. Both descriptive and normative satisficing will then be ambiopic. Hence, overall problem-solving system is ambiopic as well. This is what Fig. 4.2 depicts. The augmented agent combines both types of extreme satisficing at L_3. In consequence, overall problem-solving by this augmented agent is highly divergent and skewed, poorly fitted, and most likely dysfunctional. Digitally augmented individuals, groups, and collectives will be equally vulnerable in this way, if collaborative supervision is poor.

Non-ambiopic Augmented Metamodels

In contrast, Fig. 4.3 illustrates non-ambiopic metamodels of problem-solving. Once again, the horizontal axis shows the level of complexity of problem representation, and the vertical axis shows the complexity of solution search, both again ranging from low to high. The figure also shows two levels of processing capability. L_2 represents the processing capability of modernity as in Fig 4.2; while the greater processing

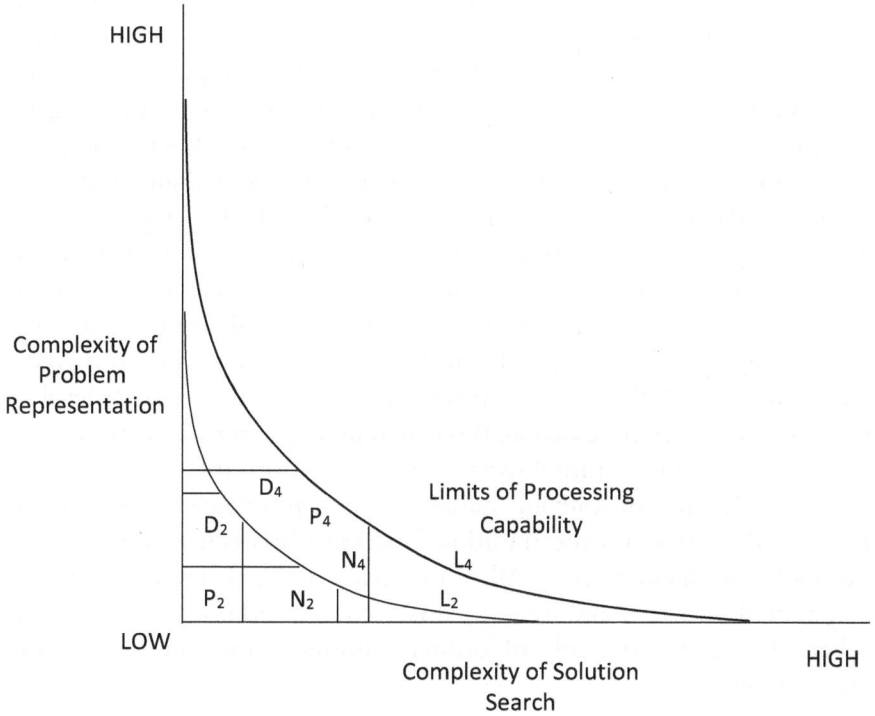

Fig. 4.3 Non-ambiopic problem-solving

capabilities of digital augmentation are here labeled L_4, to distinguish them from L_3 in Fig. 4.2. Apart from this distinction, Fig. 4.3 shares its core features with Fig. 4.2. In fact, the segments D_2, N_2, and P_2 are equivalent in both figures. They again illustrate modern, moderately assisted problem-solving and, therefore, do not require repetitive explanation.

But now consider the segment labeled D_4 in Fig. 4.3. It depicts descriptive problem-solving, given digitally augmented processing capabilities L_4. That is, seeking solutions to richly described representation of problems. However, in contrast to D_3 in Fig. 4.2, segment D_4 shows no significant divergence between the two levels of complexity. This implies the relaxation of human priors and limited artificial hyperopia. Hence, D_4 is neither ambiopic nor satisficing because it does not trade-off simplification for optimization. This scenario is non-hyperopic and non-myopic, in problem sampling and solution search. Furthermore, D_4 is shown to equal N_4. In other words, descriptive and normative methods are conflated. Neither is ambiopic nor satisficing. Instead, digitally augmented capabilities allow agents to heighten both problem representation and solution, to equal levels of complexity. By doing so, the agent mitigates myopia and hyperopia. In essence, description becomes highly computational, and normative computation is richly descriptive (Yan, 2019).

For similar reasons, the segment labeled P_4, which depicts practical maximizing, is equivalent to D_4 and N_4 as well. In fact, all three segments overlap. What this illustrates, is that the agent fully relaxes prior commitments and forgoes optimization altogether. The result is practical maximizing, highly contextual, and generative. Moreover, owing to digitally augmented capabilities, such maximizing may achieve a high level of completeness. Also note that all three metamodels intersect with a curved section of L_4. This feature illustrates a degree of possible variance, or in other words, the maximizing nature of such problem-solving. In this fashion, P_4 overcomes the traditional polarity between descriptive and normative problem-solving. All of problem-solving at level L_4 is highly augmented and non-ambiopic in this scenario, although, by the same token, P_4 shrinks the role of ordinary human intuition, values, and commitments.

Therefore, human priors are relaxed and artificial hyperopia is controlled. Both problem representation and solution search will be largely free of human myopias and excessive artificial hyperopia. This is what Fig. 4.3 depicts. Problem-solving is non-ambiopic and fully maximizing, from a practical perspective. However, as explained above, this type of problem-solving reduces the role of ordinary human intuition, values, and commitments. Granted, the system achieves greater precision and integration, but it also depletes problem-solving of important human qualities. This approach is also dysfunctional, therefore, when problem-solving warrants the inclusion of humanistic factors and commitments.

Moderately Ambiopic Augmented Metamodels

Other digitalized metamodels will be less extreme, better supervised, and moderately ambiopic. Augmented problem-solving of this kind is more balanced. It includes some human supervision of descriptive and normative satisficing, while also exploiting the benefits of augmented, practical maximizing. Agents accept a modest degree of myopia and hyperopia in sampling and search, often using both structured and unstructured data, given agreed criteria of supervision. In this kind of metamodel, the segments D_4, N_4, and P_4, will be partially distinct and not fully equivalent. The overall system of problem-solving will admit more human inputs, referencing personal and cultural values, goals, and commitments, but avoid excessive myopia, while at the same time allowing some artificial agents to operate fully independent of human supervision, but avoiding excessive hyperopia. In this fashion, augmented agents exploit digitalized capabilities, while preserving valued features of human and artificial problem-solving, thereby achieving strong metamodel fit. For this reason, many behavioral and social contexts will favor moderately ambiopic, augmented problem-solving.

4.4   Implications for Problem-Solving

Digital augmentation promises great advances for problem-solving, assuming human and artificial agents learn to function effectively as augmented agents, working together with mutual trust and empathic supervision. To some, it may seem strange to describe artificial agents in this way, almost as if they were human. Some may reject the description as fanciful, and even as dangerous. However, recent technical innovations are compelling. Artificial agents already surpass humans in many calculative functions, and recent developments enable associative and creative intelligence (Horzyk, 2016). In addition, artificial agents are rapidly acquiring empathic capability, which allows them to interpret and imitate personality, emotion, and mood. Many agents also function in a fully autonomous, self-generative fashion. When combined, these capabilities are approaching human levels, in significant respects (Goertzel, 2014), at least, to the degree required for meaningful collaboration in augmented problem-solving.

At the same time, significant challenges lie ahead, as humans respond to the rapid growth of artificial capabilities. The combinatorics are challenging. On the one hand, human absorptive capacities are limited, people habitually simplify, biases and myopias easily intrude, and learning is often truncated. While on the other hand, artificial intelligence and machine learning race ahead at unprecedented speed and scale. Indeed, we constantly see more powerful examples. However, owing to lagging skills of collaborative supervision, these technical innovations could amplify (rather than mitigate) the weaknesses of human problem-solving. Hence, humanity faces a growing challenge: to ensure that augmented problem-solving exploits the power of digitalization, while managing human needs and potential costs.

As this chapter reports, many are working on these questions. Some are optimistic (Harley et al., 2018; Woetzel et al., 2018). They point to positive developments, such as the diffusion of knowledge, greater variety of choice, and the delivery of highly intelligent services, not to mention advances in complex problem-solving. Others are more pessimistic. They highlight the contagion of digitalized falsehood, bias, and social discrimination in problem-solving, plus intrusive surveillance and manipulation, whereby elites seek to control the flow of online information and analysis (Osoba & Welser, 2017; Zuboff, 2019). The mechanisms exposed here help to explain what is occurring in these situations, namely, the deliberate use of myopia, hyperopia, and ambiopia, in digitally augmented problem-solving. Whether one is optimistic or pessimistic about the future, these mechanisms warrant urgent attention.

Myopia, Hyperopia, and Ambiopia

Among the most important topics for further research, therefore, are the risks of doing too much and too little. That is, of poorly supervised myopia plus hyperopia in sampling and search, leading to extremely divergent, ambiopic problem-solving (Baer & Kamalnath, 2017). As noted earlier, computer scientists already research similar risks. Many mitigating strategies focus on semi-supervised learning (Jordan & Mitchell, 2015). To date, however, these higher order procedures are not major topics for behavioral and organizational research (Gigerenzer & Gaissmaier, 2011). They should be. Augmented agents will confront these risks as well. Their goal will be to maximize metamodel fit in any problem context. Otherwise, augmented agents face the prospect of dysfunctional problem-solving.

The argument also highlights the role of human commitments, and especially those which serve as reference criteria about what is realistic, reasonable, and ethical, in problem-solving. Such criteria often emerge over time, are culturally embedded, and have institutional expression (Scott & Davis, 2007). Such commitments are deeply imprinted in thought and identity (Sen, 1985). For this reason, they are and often should be, difficult to change and adapt. Indeed, resilient commitments play an important role in sustaining institutions, social relations, and personalities. The risk is that absent appropriate supervision, inflexible commitments and their escalation can lead to excessive myopia and hyperopia in sampling and search. Overall problem-solving then becomes highly ambiopic for no good reason, and therefore dysfunctional.

Bounded and Unbounded

Myopic risks reflect the natural limits of human capabilities, which are widely assumed in modern thought. Whether in theories of perception, reasoning, empathy, memory, agency, or reflexive functioning, scholars assume limited human capabilities. In relation to problem-solving, Simon (1979) explains the bounded nature of human calculative rationality, and why agents satisfice against relevant performance criteria, rather than fully optimizing. As noted earlier, he formulated two broad methods of satisficing, which I label normative and descriptive. The former seeks optimum solutions for a simplified world, and the latter, satisfactory solutions for a detailed, realistic world. Simon's insights have influenced numerous fields of enquiry, including behavioral theories of problem-solving and decision-making, the management and design of organizations, and branches of economics (Gavetti et al., 2007).

However, cognitive boundedness is significantly mitigated by digital augmentation. Digital technologies massively enhance everyday processing capabilities, and especially in complex problem-solving. Humans can perceive, reason, and memorize with far greater precision, speed, and collaborative reach. At least, these extensions are now feasible. In these respects, augmented agents can be bounded and unbounded, at the same time. This occurs because human agents will likely retain their natural boundedness, especially in everyday cognitive functioning. At the same time, artificial agents will be increasingly unbounded. When both agents join in collaborative problem-solving, therefore, the resulting augmented agents could be simultaneously bounded and unbounded. In other words, they will exhibit functional ambimodality, as distinct from the organizational types of ambimodality discussed in the preceding chapter.

Satisficing then becomes more complicated, but also more important, because it can help to limit overprocessing, including the tendency toward overly hyperopic sampling and search. The role of satisficing will therefore expand and deepen. Instead of satisficing because of limited capabilities, augmented agents will satisfice because of extra capabilities. Deliberate satisficing will help to avoid unnecessary optimization. Put another way, digitally augmented agents will satisfice, not only in response to limits, but to impose limits. They will choose descriptive or normative satisficing, even when ideal optimization is feasible, or at least approachable. People sometimes do this already when they employ heuristics (Gigerenzer & Gaissmaier, 2011). Artificial agents do as well when they limit their own processing to improve speed and efficiency. Augmented agents will do the same, by managing myopia and hyperopia to maximize metamodel fit in problem-solving, forgoing possible optimization for good reasons.

This analysis has major implications for the fields mentioned earlier, which assume Simon's analysis of boundedness, including behavioral theories of problem-solving and decision-making, the management and design of organizations, and related fields of behavioral economics and choice theory. Each field will need to revisit its core assumptions, to accommodate less bounded capabilities and intentional satisficing. And when this happens, all of economics starts to look behavioral, as Thaler (2016) predicts. In similar fashion, scholars may need to rethink the assumed opacity of preference ordering, interpersonal comparison, and collective choice (Sen, 1997a). Given the expanded capabilities brought by digital augmentation, it will become feasible to seek comparative transparency, almost complete ordering, and approach optimization in some digitalized contexts. Granted, this may not be desirable. It could erode human diversity and creativity. But this type of choice will be feasible, nonetheless. Mindful of these risks, augmented humanity will need to monitor and manage the risks of over-completion in preference ordering and collective choice, and often choose to be better rather than perfect (see Bazerman, 2021).

Extended, Ecological Rationality

Another notable implication of digitalization is the extension of systematic intelligence to problem sampling and representation. In the past, everyday problems were taken as given, the intuited products of experience and sensory perception, whereas rigorous problem sampling and representation have been the preserve of empirical science. For this reason, most theories of behavioral problem-solving assume that systematic intelligence relates to solution search and selection, but rarely to problem sampling and representation. Rationality has been about finding solutions and making decisions, not about the specification of problems as such. However, digital augmentation upends these assumptions too. New tools and techniques allow augmented agents to reason systematically during problem sampling and representation. In this regard, recall the discussion of feedforward mechanisms and entrogenous mediation in Chaps. 1 and 2. Problem sampling and representation will be updated in a rapid, intra-cyclical fashion, through intelligent sensory-perception. Sampling and representation become reasoned activities. Ecological theory should therefore expand to embrace realism as well as rationality. Both aspects of problem solving will be contextual and dynamic.

Augmented agents will therefore apply intelligence to problem sampling and representation, not only to solution search and selection. For example, important problems regarding personal health, finances, and consumer preferences will be identified and curated by artificial agents, often in real time. In fact, this already happens, via smartphone applications. In the background, systems analyze and update problems in real time. However, this also entails that many processes will not be fully accessible to consciousness. In fact, as in other augmented domains, ordinary consciousness will play a different role in problem-solving. It will be an important source of humanistic guidance, but less significant as a window onto fundamental reality and truth. In all these ways, augmented problem-solving calls for an extended, ecological understanding of realism and rationality (Todd & Brighton, 2016).

This shift has another, profound implication. Important social-psychological distinctions relate to proximal versus distal processing. Construal Level Theory, for example, assumes that humans treat phenomena and problems differently, depending on their perceived spatial, temporal, social, and hypothetical distance (Trope & Liberman, 2011). If close or proximal, they are treated as more practical, short term, parochial, and risky, whereas if distal, they are more exploratory, long term, and expansive. Higgins' (1998) Regulatory Focus Theory assumes comparable distinctions. However, if digitally enabled hyperopia draws everything closer on these dimensions, then what is distal with respect to human experience and capabilities could be proximal in artificial terms. When combined, augmented agents could perceive problems as proximal and distal at the same time. This would likely lead to ambiguous or conflicting construals, and misguided sampling and search. Once again, augmented agents must learn to manage these ambiopic risks.

Culture and Collectivity

Digitally augmented problem-solving has cultural implications as well. To begin with, communities share problems which they represent and resolve at a collective level. Such problem-solving is often divided between the domains of science and technology, on the one hand, and human value and meaning, on the other. In fact, some observers of modernity refer to two dominant cultures (March, 2006; Nisbett et al., 2001). Digitalization problematizes these distinctions. Already, digitalization is transforming the creative arts and entertainment. Artificial agents make meaning and create aesthetic value. In consequence, the two cultures are blending, at least in these domains. Numerous potential benefits accrue, in terms of cultural interaction and understanding. However, it is equally possible that these trends could amplify ambiopic problem-solving and exacerbate cultural divergence and division (Kearns & Roth, 2019). Here, too, the challenges of digital augmentation are far from understood, let alone effectively supervised. It remains an open question, whether augmented agency will evolve quickly enough to manage these growing risks.

In the modern period, systematic reason and experimental science support unprecedented problem-solving capabilities. Digitalization extends this historic narrative to everyday problem representation and solution search. Yet in doing so, digital augmentation alters the dynamics of problem-solving itself. Every aspect of problem-solving becomes more intelligent and agile. However, human beings remain limited by nature and nurture, and these human factors will persist. Moving forward, therefore, research should focus on the interaction of human and artificial agents in augmented problem-solving, the novel risks of hyperopia and ambiopia, and how collaborative supervision can mitigate these risks and maximize metamodel fit. Many of these questions already loom large in computer science. They deserve equal attention from scholars in the human and decision sciences.

References

Amodeo, L., Talbi, E.-G., & Yalaoui, F. (2018). *Recent developments in meta-heuristics*. Springer.

Appiah, K. A. (2017). *As if: Idealization and ideals*. Harvard University Press.

Baer, T., & Kamalnath, V. (2017, November). Controlling machine-learning algorithms and their biases. *McKinsey Quarterly*.

Balasubramanian, N., Ye, Y., & Xu, M. (2020). Substituting human decision-making with machine learning: Implications for organizational learning. *Academy of Management Review* (online).

Bandura, A. (1991). Social cognitive theory of self-regulation. *Organizational Behavior & Human Decision Processes, 50*(2), 248–287.

Bazerman, M. (2021). *Better, Not Perfect*. Harper Business.

Bazerman, M. H., & Sezer, O. (2016). Bounded awareness: Implications for ethical decision making. *Organizational Behavior and Human Decision Processes, 136*, 95–105.

Beach, L. R., & Connolly, T. (2005). *The psychology of decision making: People in organizations* (2nd ed.). Sage.

Boussaid, I., Lepagnot, J., & Siarry, P. (2013). A survey on optimization meta-heuristics. *Information Sciences, 237*(Supplement C), 82–117.

Bruner, J. (2004). Life as narrative. *Social Research, 71*(3), 691–710.

Brusoni, S., Marengo, L., Prencipe, A., & Valente, M. (2007). The value and costs of modularity: A problem-solving perspective. *European Management Review, 4*(2), 121–132.

Burke, E. K., Gendreau, M., Hyde, M., Kendall, G., Ochoa, G., Ozcan, E., & Qu, R. (2013). Hyper-heuristics: A survey of the state of the art. *Journal of the Operational Research Society, 64*(12), 1695–1724.

Camerer, C. F. (2019). 24. Artificial intelligence and behavioral economics. In *The economics of artificial intelligence* (pp. 587–610). University of Chicago Press.

Chen, S.-H. (2017). *Agent-based computational economics: How the idea originated and where it is going*. Routledge.

Chen, J. Y. C., & Barnes, M. J. (2014). Human-agent teaming for multirobot control: A review of human factors issues. *IEEE Transactions on Human-Machine Systems, 44*(1), 13–29.

Cohen, M. D. (2006). Reading Dewey: Reflections on the study of routine. *Organization Studies, 28*(5), 773–786.

Csaszar, F. A., & Siggelkow, N. (2010). How much to copy? Determinants of effective imitation breadth. *Organization Science, 21*(3), 661–676.

Denrell, J., & March, J. G. (2001). Adaptation as information restriction: The hot stove effect. *Organization Science, 12*(5), 523–538.

Denrell, J., Liu, C., & Mens, G. L. (2017). When more selection is worse. *Strategy Science, 2*(1), 39–63.

Fiedler, K. (2012). Meta-cognitive myopia and the dilemmas of inductive-statistical inference. *Psychology of Learning & Motivation, 57,* 1–55.

Fiedler, K. (2014). From intrapsychic to ecological theories in social psychology: Outlines of a functional theory approach. *European Journal of Social Psychology, 44*(7), 657–670.

Fiedler, K., & Juslin, P. (Eds.). (2006). *Information sampling and adaptive cognition.* Cambridge University Press.

Fiedler, K., & Wanke, M. (2009). The cognitive-ecological approach to rationality in social psychology. *Social Cognition, 27*(5), 699–732.

Forster, J., Friedman, R. S., & Liberman, N. (2004). Temporal construal effects on abstract and concrete thinking: Consequences for insight and creative cognition. *Journal of Personality and Social Psychology, 87*(2), 177–189.

Gavetti, G., Levinthal, D. A., & Rivkin, J. W. (2005). Strategy making in novel and complex worlds: The power of analogy. *Strategic Management Journal, 26*(8), 691–712.

Gavetti, G., Levinthal, D., & Ocasio, W. (2007). Neo-Carnegie: The Carnegie school's past, present, and reconstructing for the future. *Organization Science, 18*(3), 523–536.

Geertz, C. (2001). *Available light.* Princeton University Press.

Gigerenzer, G. (1996). On narrow norms and vague heuristics: A reply to Kahneman and Tversky (1996). *Psychological Review, 103*(3), 592–596.

Gigerenzer, G. (2000). *Adaptive thinking: Rationality in the real world.* Oxford University Press.

Gigerenzer, G. (2008). *Gut feelings: Short cuts to better decision making.* Penguin.

Gigerenzer, G., & Gaissmaier, W. (2011). Heuristic decision making. *Annual Review of Psychology, 62,* 451–482.

Glisson, C. C. (2019). Approach to diplopia. *CONTINUUM: Lifelong Learning in Neurology, 25*(5), 1362–1375.

Goertzel, B. (2014). Artificial general intelligence: Concept, state of the art, and future prospects. *Journal of Artificial General Intelligence, 5*(1), 1–48.

Harley, D., Morgan, J., & Frith, H. (2018). *Cyberpsychology as everyday digital experience across the lifespan.* Palgrave Macmillan.

Hasselberger, W. (2019). Ethics beyond computation: Why we can't (and shouldn't) replace human moral judgment with algorithms. *Social Research: An International Quarterly, 86*(4), 977–999.

He, X., & Xu, S. (2010). *Process neural networks: Theory and applications.* Springer Science & Business Media.

Higgins, E. T. (1998). Promotion and prevention: Regulatory focus as a motivational principle. *Advances in Experimental Social Psychology, 30*, 1–46.

Higgins, E. T., & Scholer, A. A. (2009). Engaging the consumer: The science and art of the value creation process. *Journal of Consumer Psychology, 19*(2), 100–114.

Horzyk, A. (2016). Human-like knowledge engineering, generalization, and creativity in artificial neural associative systems. In *Knowledge, information and creativity support systems: Recent trends, advances and solutions* (pp. 39–51). Springer.

Jaillet, P., Jena, S. D., Ng, T. S., & Sim, M. (2016). Satisficing awakens: Models to mitigate uncertainty. *Optimization Online.*

Jenna, B. (2016). How the machine 'thinks': Understanding opacity in machine learning algorithms. *Big Data & Society, 3*(1).

Jordan, M. I., & Mitchell, T. M. (2015). Machine learning: Trends, perspectives, and prospects. *Science, 349*(6245), 255–260.

Kahneman, D., & Tversky, A. (2000). Prospect theory: An analysis of decision under risk. In D. Kahneman & A. Tversky (Eds.), *Choices, values, and frames* (pp. 17–43). Cambridge University Press.

Kahneman, D., Sibony, O., & Sunstein, C.R. (2021). *Noise: A Flaw in Human Judgment.* London: William Collins.

Kahneman, D., Rosenfield, A. M., Gandhi, L., & Blaser, T. (2016). Noise: How to overcome the high, hidden cost of inconsistent decision making. *Harvard Business Review, 94*(10), 38–46.

Kearns, M., & Roth, A. (2019). *The ethical algorithm: The science of socially aware algorithm design.* Oxford University Press.

Kitsantas, A., Baylor, A. L., & Hiller, S. E. (2019). Intelligent technologies to optimize performance: Augmenting cognitive capacity and supporting self-regulation of critical thinking skills in decision-making. *Cognitive Systems Research, 58*, 387–397.

Lauriere, J.-L. (1978). A language and a program for stating and solving combinatorial problems. *Artificial Intelligence, 10*(1), 29–127.

Lee, S. H., & Ro, Y. M. (2015). Partial matching of facial expression sequence using over-complete transition dictionary for emotion recognition. *IEEE Transactions on Affective Computing, 7*(4), 389–408.

Levinthal, D. A. (2011). A behavioral approach to strategy—What's the alternative? *Strategic Management Journal, 32*(13), 1517–1523.

Liu, C., Vlaev, I., Fang, C., Denrell, J., & Chater, N. (2017). Strategizing with biases: Making better decisions using the mindspace approach. *California Management Review, 59*(3), 135–161.

Luan, S., Reb, J., & Gigerenzer, G. (2019). Ecological rationality: Fast-and-frugal heuristics for managerial decision making under uncertainty. *Academy of Management Journal, 62*(6), 1735–1759.

March, J. G. (2006). Rationality, foolishness, and adaptive intelligence. *Strategic Management Journal, 27*(3), 201–214.

March, J. G. (2014). The two projects of microeconomics. *Industrial and Corporate Change, 23*(2), 609–612.

March, J. G., & Weil, T. (2009). *On leadership*. Wiley.

Marengo, L. (2015). Representation, search, and the evolution of routines in problem solving. *Industrial and Corporate Change, 24*(25), 951–980.

McGrath, J. L., Taekman, J. M., Dev, P., Danforth, D. R., Mohan, D., Kman, N., Crichlow, A., Bond, W. F., Riker, S., & Lemheney, A. J. (2018). Using virtual reality simulation environments to assess competence for emergency medicine learners. *Academic Emergency Medicine, 25*(2), 186–195.

Mullainathan, S., & Obermeyer, Z. (2017). Does machine learning automate moral hazard and error? *American Economic Review, 107*(5), 476–480.

Nisbett, R. E., Peng, K., Choi, I., & Norenzayan, A. (2001). Culture and systems of thoughts: Holistic versus analytic cognition. *Psychological Review, 108*(2), 291–310.

Ocasio, W. (2012). Attention to attention. *Organization Science, 22*(5), 1286–1296.

Osoba, O. A., & Welser, W. (2017). *An intelligence in our image: The risks of bias and errors in artificial intelligence*. Rand Corporation.

Ostrom, E. (2000). Collective action and the evolution of social norms. *Journal of Economic Perspectives, 14*(3), 137–158.

Pinker, S. (2018). *Enlightenment now: The case for reason, science, humanism, and progress*. Penguin.

Remeseiro, B., Barreira, N., Sanchez-Brea, L., Ramos, L., & Mosquera, A. (2018). Machine learning applied to optometry data. In *Advances in biomedical informatics* (pp. 123–160). Springer.

Scott, W. R., & Davis, G. F. (2007). *Organizations and organizing: Rational, natural and open system perspectives*. Pearson Education.

Sen, A. (1985). Goals, commitment, and identity. *Journal of Law, Economics & Organization, 1*(2), 341–355.

Sen, A. (1997a). Individual preference as the basis of social choice. In K. J. Arrow, A. Sen, & K. Suzumura (Eds.), *Social choice re-examined* (Vol. 1, pp. 15–37). St. Martin's Press.

Sen, A. (1997b). Maximization and the act of choice. *Econometrica: Journal of the Econometric Society, 65*(4), 745–779.

Sen, A. (2000). *Development as freedom.* Anchor Books.

Sen, A. (2005). Why exactly is commitment important for rationality? *Economics and Philosophy, 21*(1), 5–14.

Simon, H. A. (1959). Theories of decision-making in economics and behavioral science. *American Economic Review, 49*(3), 253.

Simon, H. A. (1979). Rational decision making in business organizations. *American Economic Review, 69*(4), 493–513.

Smith, A. (1950). *An inquiry into the nature and causes of the wealth of nations (1776).* Methuen.

Smith, V. L. (2008). *Rationality in economics: Constructivist and ecological forms.* Cambridge University Press.

Smith, A. (2010). *The theory of moral sentiments (1759).* Penguin.

Smolarz-Dudarewicz, J., Poborc-Godlewska, J., & Lesnik, H. (1980). Comparative evaluation of the usefulness of the methods of studying binocular vision for purposes of vocational guidance. *Medycyna Pracy, 31*(2), 109–114.

Stiglitz, J. E., Sen, A., & Fitoussi, J.-P. (2009). *Report of the commission on the measurement of economic performance and social progress.*

Thaler, R. H. (2000). From homo economicus to homo sapiens. *Journal of Economic Perspectives, 14*(1), 133–141.

Thaler, R. H. (2016). Behavioral economics: Past, present, and future. *American Economic Review, 106*(7), 1577–1600.

Todd, P. M., & Brighton, H. (2016). Building the theory of ecological rationality. *Minds and Machines, 26*(1), 9–30.

Trope, Y., & Liberman, N. (2011). Construal level theory. In P. A. M. V. Lange, A. W. Kruglanski, & E. T. Higgins (Eds.), *Handbook of theories of social psychology* (Vol. 1, pp. 118–134).

Woetzel, J., Remes, J., Boland, B., Lv, K., Sinha, S., Strube, G., Means, J., Law, J., Cadena, A., & von der Tann, V. (2018). *Smart cities: Digital solutions for a more livable future.* McKinsey Global Institute.

Yan, H. (2019). *Description approaches and automated generalization algorithms for groups of map objects.* Springer.

Zuboff, S. (2019). *The age of surveillance capitalism: The fight for a human future at the new frontier of power.* Profile Books.

5

Cognitive Empathy

At the dawn of European Enlightenment, Descartes (1998) meditated on his own conscious life and concluded *cogito ergo sum*, meaning "I think therefore I am." He thereby located the core of selfhood in reasoned, reflexive thought. Departing from premodern assumptions, he accorded supernatural forces a minor, ancillary role. For Descartes, and for many who followed him, the exercise of autonomous, intelligent agency was the distinguishing feature of being human, not the replication of some mythical narrative or religious ideal. The experience and explanation of reflexive selfhood were transformed. Understanding of intersubjectivity was equally affected. To understand another person, one must empathize with her or his reasoning and its relationship to the person's speech and action. For this reason, advocates of the Enlightenment look for realism and rationality in other minds and are dissatisfied with superstition and rituals (Pinker, 2018).

Modernity therefore celebrates intelligent, autonomous agency, and assumes that people can and should empathize at this level. This implies that other minds are potentially accessible and explicable to consciousness. Many social and behavioral sciences share this outlook. For example, theories of institutions invoke empathy with other minds to explain

P. T. Bryant, *Augmented Humanity*, https://doi.org/10.1007/978-3-030-76445-6_5

collective logics and decision-making (Thornton et al., 2012). For philosophers, it leads to the problem of other minds and how to interpret them (Dennett, 2017). Contemporary psychologists also research intersubjectivity, and especially cognitive empathizing, which is the process whereby individuals represent and comprehend the thoughts and reasoning of others, or read other minds (Decety & Yoder, 2016; Schnell et al., 2011). Likewise, assumptions about cognitive empathy and its limits are central to modern theories of ethics and justice (Sen, 2009), as well as experimental microeconomics (Singer & Fehr, 2005). All these fields recognize the significance of cognitive empathizing, that is, the representation and solving of problems of other minds.

Furthermore, modernity exhibits a steady stream of technological innovations which support cognitive empathizing. Earlier generations exploited the telegraph and telephone, which greatly expanded understanding of other people's thoughts. More recently, digital technologies, including artificial intelligence and ubiquitous online services, allow people to learn more and more about others' beliefs, reasons, and mental worlds (Wulf et al., 2017). Examples proliferate. Social networks show who and what is liked; using their smartphones, people can share experiences instantaneously on a global scale; connected devices enable the real-time sharing of ideas and opinions; data about online behavior are then used to predict personal preferences; virtual assistants such as Apple's Siri and Amazon's Alexa (note the humanizing names) mediate communication like actual persons. Additional capabilities are now emerging, including artificial personality and affective computing, wearable and potentially implantable devices, which will enable the digitalized interpretation and imitation of human mood and emotion (Poria et al., 2017).

In fact, some artificial agents can already interpret and imitate significant aspects of human facial expression, empathy, and personality. As noted previously, in recent experiments of customer service by telephone, callers could not distinguish between human and artificial agents (Leviathan & Matias, 2018). The artificial agent sounded fully human, in terms of its expressions and empathy. Granted, such innovations are nascent and too often flawed, but they will improve and become ubiquitous. Digital assistants will mediate significant aspects of intersubjectivity and cognitive empathizing. By exploiting such innovations, people can aspire to deeper understanding of each other and themselves. Subjectivity and self-consciousness could be digitally augmented as well. Digitalization will transform cognitive empathizing with others and the self. But this process will take time and new regulatory systems are needed. In fact, in the short term, radically augmented, cognitive empathizing could overwhelm many people and destabilize both personalities and communities (Chimirri & Schraube, 2019). Most people are not equipped to manage highly transparent minds, even with the support of artificial intelligence. People do well to intuit what is hidden and often unformed in other minds, and in their own (Davidsen & Fosgerau, 2015). Apart from anything else, mental noise and nonsense would drown out much of the signal. Empathy can also be emotionally exhausting, at the best of times (Bandura, 2002).

Therefore, human empathic capabilities are likely to remain strictly limited, at least for the foreseeable future. As in other forms of complex problem-solving, habitual and routine procedures will often take precedence. Similarly, biases and myopias will likely continue as well. For these reasons, as in problem-solving generally, digitalization may compound, rather than ameliorate, the traditional dilemmas of cognitive empathizing (Mullainathan & Obermeyer, 2017). For example, if racial and gender biases are encoded into the sampling and representation of others minds, and then carried over into the training of algorithms, digitalization amplifies discriminatory judgments (Noble, 2018). Too many examples already exist, of poorly supervised, biased machine learning, imputing erroneous states of mind to racial or gender groups (Eubanks, 2018; Osoba & Welser, 2017). Comparable biases fuel the febrile tribalism of online xenophobia, reinforcing the perceived deviance and irrationality of others.

Yet bias and myopia are not the only dilemmas. As digitally augmented capabilities become more powerful and ubiquitous, augmented agents can err in the opposite direction, employing these capabilities to sample and search too widely, thereby over-sampling and over-searching other minds. Cognitive empathizing could go too far in these respects. Agents might gather too much information about other minds and apply overly complex algorithms to interpret them. In other words, the hyperopic risks of problem-solving which were examined in the preceding chapter, also impact cognitive empathizing, when it is conceived as solving problems of other minds. Recent studies already demonstrate these effects.

They show the negative consequences of hyperopic sampling and search in obsessive cognitive empathizing, especially when combined with persistent myopias (Lemaitre et al., 2017). Reconsider an example given earlier. Some machine learning agents are trained using racially biased data. The agent then over-samples and over-searches the utterances and behaviors of others, guided by biased supervision. In doing so, it gathers vast amounts of evidence to reinforce the erroneous priors, in consequence, automating framing and confirmation biases at scale (Baer & Kamalnath, 2017). This results in overly ambiopic cognitive empathizing, defined as the combination of divergent degrees of simplification and complexity in problem-solving about other minds.

We need to ask, therefore, under which conditions will digitalization enable more effective cognitive empathizing, widening the appreciation of others' thoughts, beliefs, and reasons, rather than perpetuating and compounding erroneous myopias and biases or amplifying noise; and which additional procedures might help to reduce risks for cognitive empathy, while enjoying the potential benefits of digital augmentation? Moreover, these questions are increasingly urgent (Bolino & Grant, 2016). Organizations and groups all rely on cognitive empathy. However, evidence suggests that the speed and scale of digitalization are outpacing ordinary empathic capabilities (Mullainathan & Obermeyer, 2017). Significant aspects of digitally augmented mentality, both individual and collective, are eluding self-supervision. And myopias and biases are persistent. Not surprisingly, many people already suffer from digitally distorted, cognitive empathizing which leads to misunderstanding and mistrust.

In contrast, artificial agents can sample and search other minds with unprecedented speed and power. Through social networks, messaging applications, and the like, billions of people share extraordinary details of their thoughts, feelings, and personal lives, which only artificial agents have the capability to analyze and aggregate. When these systems combine with ordinary humans, however, the results can be highly ambiopic cognitive empathizing: overly distal and complex in some respects (owing to artificial hyperopia), yet overly proximal and simplified in other ways (owing to human myopia). If this occurs, agents of any modality—whether individuals, groups, or collectives—will tend to misconstrue other's positions and perspectives and be prone to misjudgments and attribution errors. As a result, some other minds will be unfairly perceived as unrealistic, irrational, or deviant, while others will be misperceived as fully rational and realistic. Indeed, studies show that cognitive empathy is already degrading in some digitalized domains, and arguably for these reasons (Miranda et al., 2016).

The digital augmentation of cognitive empathy therefore poses major opportunities and risks for humanity. If well supervised, augmented empathizing could enhance mutual understanding, trust, and cooperation. But if ambiopic tendencies are left unchecked, digitally augmented empathizing can skew in a few directions and be overly divergent or convergent. First, if human myopias and biases are encoded into augmented agents, and then amplified by hyperopic processing, the resulting divergence will erode cognitive empathy, heighten mistrust, and fray the coherence of collective mind. Second, if artificial agents dominate cognitive empathizing, they could smother ordinary human intuition and instinct and erode the diversity and delight of human relating. Alternatively, third, if human agents dominate, they could impose myopias and biases which stifle and distort the sampling and search of other minds. The current chapter examines these challenges. As a first step, we need to examine the core mechanisms of cognitive empathizing more deeply.

5.1   Theories of Cognitive Empathy

Jerome Bruner (1996) predicted a generation ago, that the next chapter of psychological research would focus increasingly on intersubjectivity, being the mechanisms by which people appreciate others' subjective experience of self and the world. As earlier sections of this chapter suggest, his prediction has proven correct. In particular, contemporary psychologists investigate cognitive empathy, defined as reading the thoughts and reasons of others (Liljenfors & Lundh, 2015). Moreover, as Descartes' (1998) meditations illustrate, people also cognitively empathize with themselves, whereby they form a sense of self as a reasoning agent. In fact, acquiring this reflexive capability is an important phase of child development, along with the capability to distinguish other minds as separate from one's own (Katznelson, 2014). Cognitive empathy is therefore critical to a range of psychological and developmental processes.

Psychology of Cognitive Empathy

The core psychological mechanism of cognitive empathy is mentalization, which is the process whereby persons apprehend and form mental representations, of their own and others' mental states (Fonagy & Campbell, 2016). Mentalization is both internally and externally focused, on self and others, respectively. It also encompasses affective states, can be explicit and effortful, or implicit and relatively effortless, as a habit of mind (Liljenfors & Lundh, 2015). Notably, the construct of mentalization is well established in numerous fields. It is the subject of extensive research in clinical and cognitive psychology (Guerini et al., 2015), neuroscience (Schnell et al., 2011), education and learning (Haake et al., 2015), and now affective computing (Varga et al., 2018). Not surprisingly, mentalization also has neurological correlates (Ferrari & Coude, 2018). Moreover, because mentalization enables the perception and comprehension of cognitive states, it encompasses aspects of metacognition as well (Lindeman-Viitasalo & Lipsanen, 2017). Given this connection, its relevance will predictably spread to other fields, including management and professional studies, in which empathy receives increasing attention (Orbell & Verplanken, 2010; Ze et al., 2014). In summary, mentalization is central to the perception of, and empathy with, the subjective life of self and others.

Two types of mentalization support cognitive empathy with other minds. First, there is explicit, external, cognitive mentalization, that is, the deliberate attempt to represent and understand others' cognitive states, including the categories, concepts, beliefs, and logics which others employ in reasoning. Notably, this type of cognitive empathy is often implicit in classical theories of decision-making and formal problem-solving (March, 2014). Such theories assume that it is possible to observe and assess the reasoning of others, albeit about simplified problems and choices. Second, there is implicit, external cognitive mentalization, being the intuitive, simplified representation and comprehension of others' cognitive states. This type of cognitive empathizing is implicit within descriptive, behavioral, and informal theories of problem-solving. These theories assume that it may be impossible, and sometimes unnecessary, fully to comprehend the thoughts and reasoning of others. For a start, others' motivations and commitments can be opaque, and hard to deter-mine. In fact, as Chap. 1 explains, mounting evidence suggests that deeper states of mind are not directly accessible to consciousness, and even less so, with respect to other minds. Emotional states also play a role, and they constantly wax and wane. In addition, other persons often use informal, heuristic strategies when making choices and decisions; which is to say that human minds are often murky or muddled and resist interpretation.

Consequently, like other cognitive functions—including reasoning and attention—the capability for mentalization is limited. In addition, information about other minds is often incomplete, difficult to gather and organize, or simply inaccessible. And because other persons inhabit a plurality of mental worlds, with different positions and points of view, mentalization is further constrained by positional ambiguity (Sen, 1993). Hence, people cannot clearly identify another person's point of view. For all these reasons, mentalization is often approximating. The best one can hope for is to read other minds in a way that is no worse, and no less empathic, than other plausible readings. Indeed, people consistently maximize in this fashion, without negative consequences (Schneider & Low, 2016). Especially within shared cultures, they take much for granted, and reliably intuit each other's mental states. Although, it is important to note that major deficits in mentalization can be symptom-atic of clinical disorder (Dimaggio & Lysaker, 2015).

In any case, effortful, precise mentalization is often unnecessary or ineffective. For example, when people are engaged in purely procedural action, there may be no need to grasp the exact details of others' beliefs and reasoning. Effortful mentalization could even interrupt the flow of collective thought and action. In fact, a degree of empathic opacity is inherent and even helpful to collective mind. Such opacity invites mutual trust and civility and avoids the dilemmas and discomfort of empathic transparency (Sen, 2017). Indeed, cognitive opacity is often preferable to revealed disorder or deception, about others and oneself. Granted, there are occasions when mentalization needs to be heedful and effortful, striv-ing for precision and transparency (Weick & Roberts, 1993). Agents then upregulate mentalizing functions. But when it can be appropriately imprecise, people downregulate mentalization and simplify in cognitive empathizing. They do so naturally, when capabilities are stretched to the limit, or because habitual and routine mentalization are sufficient to secure desired outcomes (Wood & Rünger, 2016).

Empathic Satisficing

Other minds are therefore complex and often hard to read. In fact, trying to understand other minds is a type of complex problem-solving. And as philosophers attest, the problem of other minds is a wicked one (Parfit, 1984). Empathic capabilities are limited, and trade-offs are frequent, as agents simplify the representation and solution of other minds. As in problem-solving more generally, therefore, people simplify in cognitive empathizing (Baker et al., 2017; Polezzi et al., 2008). In this sense, they cognitively "empathice" about other minds, as a species of satisficing in complex problem-solving. They accept simpler empathicing outcomes, rather than seeking optimal empathizing ones.

Like other forms of satisficing, this results in two major patterns of cognitive empathicing. To quote Simon (1979, p. 498) again: "decision makers can satisfice either by finding optimum solutions for a simplified world, or by finding satisfactory solutions for a more realistic world." The same distinction applies to cognitive empathicing when we view it as a type of complex problem-solving. First, people can simplify the representation of other minds, and seek optimal solutions about them. This results in normative cognitive empathicing, which seeks to optimize solutions. People then look for deliberate calculative reasoning in others. As an example, consider the analysis of preferential choice within classically inspired microeconomics. Theories of this kind assume: (a) that agents are uniformly self-interested and rational; (b) that most agents prioritize utility maximization; and (c) that choices are made by persons in a rational, calculative fashion. In other words, the problems of other minds are simplified, so that solutions can be optimized.

Second, people describe more complex, realistic problems of other minds, and seek satisfactory solutions. This generates descriptive cognitive empathicing, which prioritizes the realistic representation of other minds. To illustrate this approach, consider the analysis of preferential choice in behavioral economics. Such theories assume that: (a) people are influenced by a wide range of factors, including beliefs, emotions, motivations, and commitments; (b) that agents prioritize a range of process and outcome conditions, including utility; and (c) they make decisions using various principles and logics. From this perspective, other minds are expressions of *Homo sapiens*, rather than *Homo economicus* (Thaler, 2000). Hence, it requires more resources to process representational complexity. To summarize, in descriptive cognitive empathicing, the problems of other minds are more realistic and relatively complex, and solutions are therefore satisfactory, rather than optimal.

Furthermore, implicit cognitive empathicing is an important mechanism of routine and collective mind, which emerge as people mentalize in purposive action (see Becchio et al., 2012; Schneider & Low, 2016). Through implicit cognitive empathicing, that is, groups of people intuit each other's thinking and develop an effortless appreciation of their common beliefs and patterns of reasoning. People thereby attribute comparable mental states and processes to each other. Granted, these representations are imprecise, as a form of implicit mentalization. But they are frequently reliable enough to maintain procedural thought and action. In this way, routines of cognitive empathicing mediate collective mind and choice (see Sutton, 2008). Moreover, as Chap. 3 explains, these characteristics of collectivity do not require the aggregation of individuals' more complex, mental processes and states (see Zhu & Li, 2017). Collective mind and choice are not aggregation puzzles. As in other scenarios of routinization, the problem of aggregation fades away, replaced by the downregulation of individual cognitive differences, and the upregulation of shared patterns of cognition.

Practical Empathicing

To manage within their constraints, human agents therefore develop practical, heuristic methods of representing and solving other minds. For example, they rely on cultural signs and symbols and use these to infer others' mental states and world (Morris et al., 2015). Going further, businesses use information about people's demographic and consumption patterns to predict their future preferences. Sporting teams assume that the opposition knows the rules of the game and will probably adhere to them. Other things being equal, practical empathicing assumes that other minds are realistic, reasoning, and self-regulated, at least to the degree required for organized social life (Bandura, 2002). Although when human emotion and idiosyncrasy intrude, anything might happen.

Modern systems of justice exhibit similar patterns. Courts and juries review information about agents' actions and utterances, predicated on the assumption that people are reasoning agents, and that it is possible to infer and assess their cognitive states and processes. In contrast, demonstrable mental illness or deficits can be a defense. To formalize all of this, some scholars argue that, in principle, it is possible to optimize cognitive empathy about others' motives and reasons. This leads to theories of justice which assume universal principles of optimal cognitive empathy. John Rawls (2001) broadly supports this position. He believes it is possible to attain a view from everywhere, *sub specie aeternitatis*, at least about fundamental features of other minds. Others disagree. Amartya Sen (2009), for example, argues that cognitive empathy is consistently and inherently incomplete. Other minds are forever partially opaque or translucent. From this perspective, justice is deeply contextual, informed by cultural context, position, and commitments. That said, both perspectives assume that agents can be understood as reasoning and self-regulated, at least to a significant degree. They disagree about the limits of cognitive empathy in these contexts and, hence, about how much is accessible to external mentalization.

Cognitive empathicing is equally important for civic and political institutions. Via empathicing, communities build consensus, by exchanging ideas and debating policies and principles of governance (Scanlon, 1998). This entails the widespread exchange of opinion, which is common to contractarian and communitarian perspectives. All are modern political visions, in this respect, because they acknowledge the importance of reasoned intersubjectivity. Cognitive empathizing is therefore critical to the functioning of such systems. It allows people to recognize the intelligent agency of others and sustain a sense of collective mind. Liberal democracy is certainly reliant on these mechanisms, and hence vulnerable to their disruption (Bandura, 2006). Not surprisingly, therefore, autocrats often try to subvert these processes. They try to control rather than liberate cognitive empathy. Although, in these respects, autocracies are modern too, because they also recognize the force of collective mind and then try to stifle it, perhaps by cultivating "false consciousness," as Marx and Engels once argued (Kołakowski & Falla, 1978). Some worry that digitalization heightens this risk, by giving more power to power (Helbing et al., 2019).

Figure 5.1 summarizes the resulting metamodels of cognitive empathizing. It mirrors the analysis of general problem-solving shown in Fig. 4.1 in Chap. 4. The new figure shows the two major components of cognitive empathizing: the representation of problems of other minds, and the solutions to such problems. For both dimensions, the figure shows their complexity as high or low. First, quadrant 1 summarizes the ideal, optimizing metamodel of cognitive empathizing, consisting of highly complex, best solutions to highly complex, best representations of other minds. Hence, I use the term empathizing here, rather than empathicing. Quadrant 2 then shows descriptive, cognitive empathicing, consisting of less complex, satisfactory solutions to complex representations of other minds. Next, quadrant 3 summarizes normative, cognitive

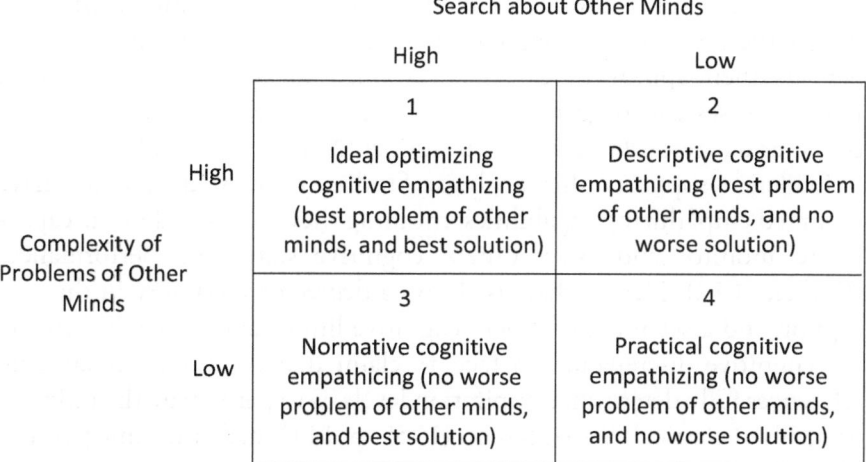

Fig. 5.1 Metamodels of cognitive empathizing

empathicing, which is complex solutions to simplified representations of other minds. And quadrant 4 shows practical cognitive empathicing, which is finding satisfactory solutions to simplified problems of other minds. This metamodel is therefore another type of empathizing, neither optimizing nor empathicing.

Empathizing and Discrepancy

It is also important to note, that the object of cognitive empathicing is frequently a form of satisficing itself. That is, people empathice in understanding other's satisficing. Put another way, people use empathicing heuristics, to represent and solve other's cognitive heuristics. Once again, this is common in cultural interactions, in which people often rely on cognitive shortcuts to read other minds (Henrich et al., 2001). It follows, therefore, that cognitive empathicing admits a range of mental performances by others, rather than fixed patterns of belief and reasoning. Empathicing agents are frequently insensitive, therefore, to cognitive variance in others. They neither perceive nor evaluate discrepant reasoning in a determinant fashion (see Wood & Rünger, 2016). Cognitive empathicing thus maximizes and people grant each other mental slack. That said, when other's performances fall below acceptable minima, cognitive empathicing will trigger the perception of significant discrepancy in other minds. Other persons then appear unrealistic, irrational, or deviant in some way. Extreme cases can trigger cognitive antipathy, meaning others are perceived as dangerously deviant or irrational (Nath & Sahu, 2017).

Evidence supports this analysis. Studies show that humans have bounded empathizing capabilities, meaning they are limited in the capacity to monitor and assess others' cognitive states and performances (Fiedler, 2012). Hence, there is always a degree of variability in the perception and assessment of other's cognitive limits, and hence, in perceiving cognitive discrepancy. In fact, to claim that cognitive empathicing references fully determinate aspiration levels, is to perpetuate the rationalist ideals of classical theory. Simon (1955, p. 111) made the same point in his original, groundbreaking exposition of aspiration and satisficing: "…there are certain dynamic considerations, having a good psychological foundation … as the individual, in his exploration of alternatives, finds it *easy* to discover satisfactory alternatives, his aspiration level rises; as he finds it *difficult* to discover satisfactory alternatives, his aspiration level falls." The same dynamic process is found in cognitive empathicing.

Granted, some instances of cognitive empathizing approach full determination and fixed aspiration levels, especially when reasoning must be highly systematic. For example, in some highly technical domains—such as the piloting of aircraft—we hope that the responsible agents think clearly and interpret each other almost perfectly. To be sure, classical theories aspire in this direction. Recall the earlier analysis of classical microeconomics. But these situations are an important subclass of the problems of other minds, not the whole universe (Sen, 1997). Many forms of cognitive empathizing are empathicing, entailing a range of satisfactory performances and, consequently, a degree of insensitivity to variance. Empathic translucence, or partial transparency, is often appropriate and effective.

Digitalization of Cognitive Empathizing

Turning next to the impact of digitalization on these processes. As noted previously, digital innovations allow people to share their mental lives in real time, on a global scale, even if much remains opaque. In addition, newer digital technologies can simulate human expression and emotion. Affective computing and artificial personality will soon be commonplace (Poria et al., 2017). In fact, the augmentation of empathy and personality is now a major field of computer engineering, already finding applications in education, automobiles, the office, and home. For all these reasons, digitalization will transform cognitive empathizing. Over time, augmented agents will be capable of richly descriptive, rapid cognitive empathizing, at every level of agentic modality and mind. Even so, natural human limitations will persist, and augmented agents will often inherit cultural stereotypes and biases about other minds.

Hence, divergence can occur in cognitive empathicing, as in other areas of augmented problem-solving. Humans may remain myopic and not sample or search other minds far enough, while artificial agents may be hyperopic and sample and search other minds too extensively. Moreover, each type of agent could easily reinforce the inherent tendencies of the other. The overall result will be divergent patterns of oversimplification and over-complexity, in the representation and solution of other minds. In short, cognitive empathicing will be highly ambiopic (Liu et al., 2017). Alternatively, the system could be overly convergent. Artificial components might overwhelm the human, or vice versa. Whether by default or design, cognitive empathicing could be hijacked by artificial or human agency.

Summary of Augmented Cognitive Empathizing

Based on the preceding discussion, we can summarize the features of digitally augmented, cognitive empathicing conceived as a type of complex problem-solving. First, empathicing integrates two main processes: the sampling and representation of problems of other minds, and the search for solutions to such problems. Second, these systems often include encoded cultural and other commitments, which guide sampling and search. Third, much cognitive empathicing is empathicing and vulnerable to overly myopic and hyperopic tendencies, resulting in highly ambiopic outcomes. And if this occurs, other minds are more likely to appear unrealistic, irrational, or deviant, and hence less trustworthy. Fourth, implicit cognitive empathicing is central to mental routine and collective mind, which do not require any process of aggregation. Fifth, agents are partially insensitive to variance in others' reasoning, given that empathicing simplifies and approximates. Many of these topics are already foci of research in computer science, for example, in sentiment analysis and artificial personality (Amodeo et al., 2018; Burke et al., 2013). What is not yet adequately understood is how they will impact cognitive empathicing in augmented agency and especially comprehension and trust between human and artificial agents, which are critical for their collaborative supervision.

5.2   Metamodels of Cognitive Empathizing

This section illustrates representative metamodels of augmented, cognitive empathicing. The illustrations adapt the earlier analysis of augmented problem-solving in the preceding chapter. This reflects the fact that cognitive empathizing can be understood as a type of complex problem-solving. Hence, the figures presented below are like those in Chap. 4, although the new figures differ in one obvious, critical respect. Rather than illustrating problem-solving in general, the figures will illustrate cognitive empathizing, including the depiction of empathicing. Also, the following discussion focuses primarily on digitally augmented mentalizing capabilities at level L_3, rather than the lesser capability level L_2.

Highly Ambiopic Empathizing

Some augmented cognitive empathizing is highly ambiopic. This will be the case when the representation and solution of other minds are very divergent, in terms of their relative complexity or simplification. Figure 5.2 illustrates this type of system, adapting Fig. 4.3 from Chap. 4. In the new figure, the axes show the two major activities of empathizing, being the representation of problems of other minds, and the search for solutions to such problems. Each dimension ranges from low to high complexity. The figures also depict the natural limits of modern, moderately assisted mentalizing capabilities, which are labeled L_2 and the higher level of digitally augmented mentalizing capabilities labeled L_3. These capabilities again reach limiting asymptotes of high complexity for both problem representation and solution search.

Figure 5.2 depicts alternative metamodels of cognitive empathicing, which optimize on one dimension and simplify on the other. First, the metamodels labeled N_2 and N_3 are normative cognitive empathicing, exploiting augmented mentalizing capabilities at levels L_2 and L_3, respectively. These metamodels combine optimizing solutions about relatively simplified problems of other minds. This implies myopic problem representation, plus hyperopic solution search. Notably, problem representation does not change between N_2 and N_3, which suggests the persistence of human sampling myopia in these metamodels. Prior simplifications persist, that is, in the representation of other minds, despite the increase in capabilities at L_3. This results in extreme normative empathicing. Granted, this could sometimes be appropriate, especially when others' cognitions relate to core human values, commitments, or cultural

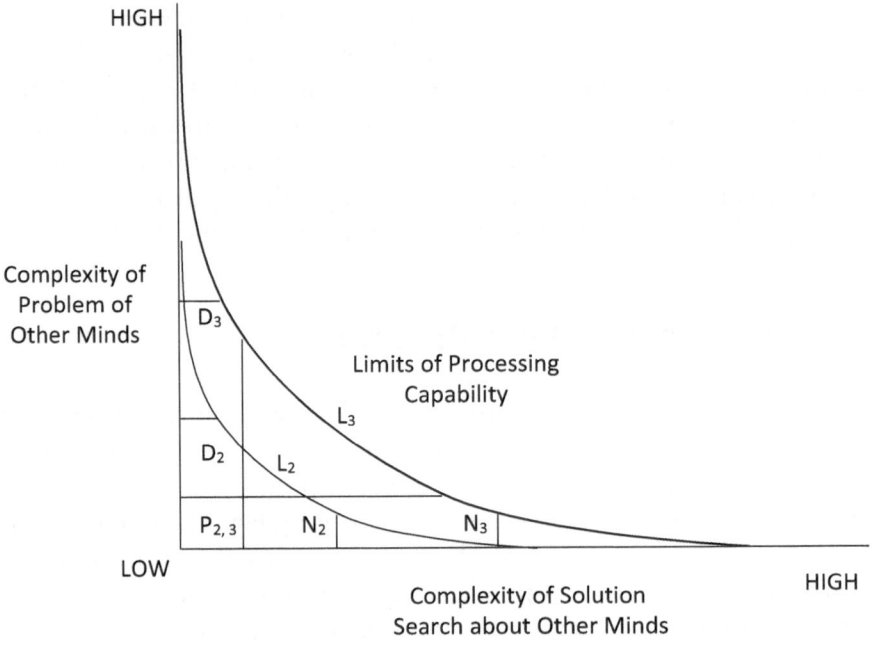

Fig. 5.2 Highly ambiopic empathicing

norms. Finally, there is a range of empathicing models, shown by the curved intersections of N_2 with L_2 and N_3 with L_3. All options along these intersections are no worse than each other and hence can be maximizing.

Second, the metamodels labeled D_2 and D_3 indicate descriptive cognitive empathicing, again exploiting mentalizing capabilities at L_2 and L_3, respectively. They combine complex problems of other minds, with simpler, satisfactory solutions. This implies hyperopic sampling and problem representation, with myopic solution search. Importantly, the solutions of other minds in D_2, are translated to D_3 without any increase in complexity. Priors persist in solution search, despite the increase in capabilities. This results in extreme descriptive empathicing. For example, consider the digitalization of collective choice, in which hyperopic sampling results in complex problem representations, which are resolved using simple choice procedures. Once again, a range of possible empathicing models is shown by the curved intersections of L_2 with D_2, of L_3 with D_3. It is also notable, that N_3 and D_3 only partially overlap in practical empathizing P_3. Furthermore, owing to the persistence of myopic commitments, these options have not expanded and P_2 is equivalent to P_3. This may be adequate for everyday life, but neither problem representation nor solution search is well specified.

Now assume a poorly supervised, augmented agent which combines N_3 and D_3. Myopic human priors remain entrenched, and artificial hyperopia is largely unchecked. Hence, both the representation and solution of problems of other minds will be anchored in human myopias. At the same time, artificial processing is highly hyperopic. In these situations, there are two potential patterns of distortion. Either augmented agents will adopt overly simplified explanations of overly complex representations of other minds, that is, extreme descriptive cognitive empathicing (D_3), or alternatively, they will adopt overly complex explanations of overly simplified representations of other minds, that is, extreme normative cognitive empathicing (N_3). Moreover, sometimes they might do both at the same time. In all scenarios, agents are more likely to perceive other minds as discrepant, deviant, and irrational. We already see such effects in the rise of online antipathy between different groups, where artificial systems reinforce and amplify encoded biases. Critical questions therefore arise: how can augmented agents supervise the retention or relaxation of human commitments in cognitive empathizing; relatedly, how can they manage the risks of myopia and hyperopia in these contexts, and maximize metamodel fit; and finally, how can digitalization enhance cognitive empathy and trust?

Non-ambiopic Empathizing

Other agents will be non-ambiopic in cognitive empathizing. Problem representation and solution search will exhibit comparable degrees of complexity. This implies that both myopia and hyperopia are relatively low, and the agent is balanced in this respect. The result is a type of practical empathizing, in which problem representation and solution search are of comparable complexity. Indeed, digitalization makes this kind of

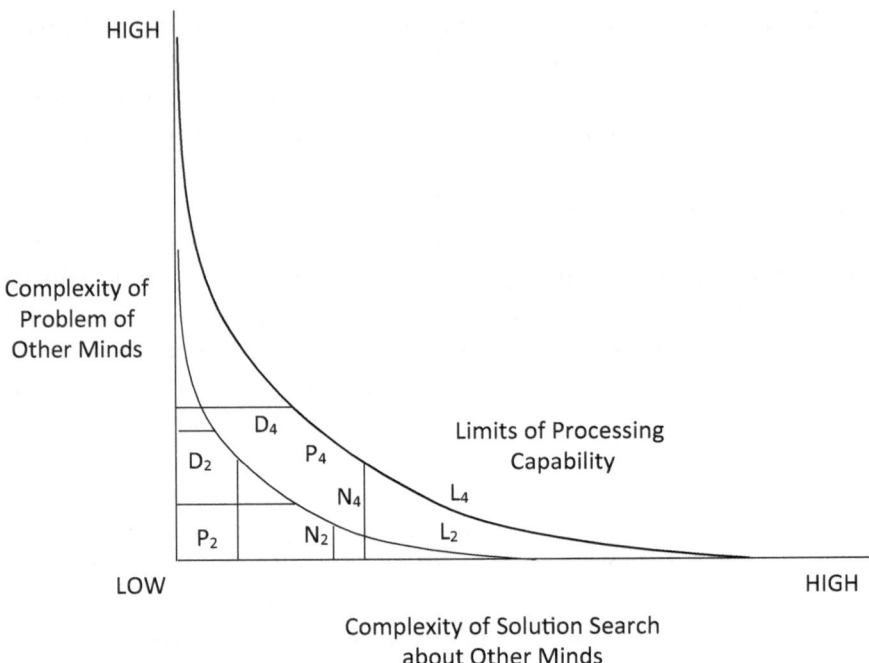

Fig. 5.3 Non-ambiopic empathizing

empathizing increasingly feasible, because digital augmentation enables detailed, precise, and intelligent sampling of other minds, combined with rigorous solution search. For the same reason, there will be fewer trade-offs. Figure 5.3 illustrates this type of metamodel, building on the illustration of non-ambiopic problem-solving in Fig. 4.3. Once again, axes represent two major components of cognitive empathizing, with two levels of mentalizing capabilities, labeled L_2 and the higher level L_4. The metamodels at level L_2 are the same as the previous figure, and therefore do not warrant repeated description.

Notably, the metamodels labeled D_4 and N_4 are both non-ambiopic. Hence, neither is clearly descriptive nor normative, in the classic sense. Rather, both are equivalent to practical empathizing P_4. In fact, all three metamodels overlap. Cognitive empathizing has conflated into the same set of digitally augmented processes. Such metamodels have been very unusual in ordinary human empathizing, although they are observed among scientists and some expert professionals, who form deep, clear understanding of each other's thoughts and intentions, often doing so with the support of sophisticated technologies. Indeed, a primary goal of the scientific method is to reduce subjective variance and instill precision into the reading of other minds. In these contexts, low ambiopia is appropriate.

Any agents who empathize in this fashion, therefore, will likely view each other as fully realistic, rational, transparent, and trustworthy. By the same token, however, this kind of empathizing will homogenize other minds, by making agents highly commensurable to each other. Some may welcome this development, hoping for better understanding of other minds. It will certainly help in complex, technical task domains. Yet others will regret the trend, concerned that digital augmentation will smother intersubjective diversity and intuition. Indeed, as noted previously, empathic ambiguity and a degree of opacity or translucence are valued in some contexts, especially in creative artistic and innovative pursuits (March, 2006). In these respects, the conflation of D_4, N_4, and P_4 represents the radical transformation mentioned earlier. Subjectivity itself has been fully augmented by digitalized perception, thought, and feeling. As a result, however, ordinary intuition and instinct are bleached.

Summary of Cognitive Empathizing

In summary, there are reasons to hope and be wary. Fully digitalized empathizing (shown in Fig. 5.3) could be accurate and transparent, but homogenized and lack human intuition and instinct. Whereas humanized empathizing (shown in Fig. 5.2) will be diverse and intuitive, but translucent and less accurate. Figure 5.4 summarizes the overall patterns. The vertical dimension shows the level of cognitive empathy versus antipathy. The horizontal dimension shows the degree of ambiopia in cognitive empathizing, moving toward extreme descriptive empathicing on the left and toward extreme normative empathicing on the right. Six patterns of cognitive empathizing are positioned on these dimensions.

First, the figure shows the two modern metamodels in Figs. 5.2 and 5.3 (labeled D_2 and N_2), which are moderately ambiopic, descriptive and normative empathicing, respectively. Both exhibit moderate cognitive

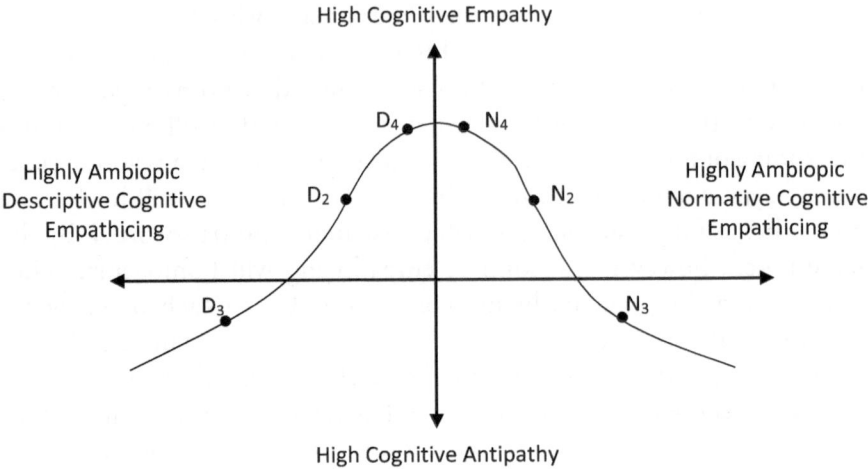

Fig. 5.4 Summary of cognitive empathicing

empathy. Second, the figure also shows the two metamodels in Fig. 5.2 (D_3 and N_3), which represent highly ambiopic, extreme descriptive and normative empathicing, respectively. In fact, both systems exhibit moderate cognitive antipathy. Third, Fig. 5.4 positions the two metamodels in Fig. 5.3 (D_4 and N_4), which represent non-ambiopic descriptive and normative patterns of empathizing, respectively. Both metamodels exhibit high potential for cognitive empathy, although, as noted earlier, the risk is less diverse and intuitive empathizing. In summary, there are three broad patterns in Fig. 5.4, from high cognitive transparency and empathy, to balanced cognitive translucence and empathy, to cognitive opacity and antipathy. And depending on the context, each could be a good fit, even antipathy sometimes. However, moderate ambiopia and translucence will often be most advantageous, especially in ordinary life, because they foster reliable cognitive empathicing while also allowing for diversity and intuition.

5.3 Wider Implications

For good and ill, digitalization is transforming cognitive empathy. Digital innovations are rapidly augmenting the capability to perceive and read other minds. Yet human limitations and biases persist. As in all complex problem-solving, therefore, augmented agents will simplify when reading other minds. They will cognitively empathice. Two major patterns of distortion are most likely. First, augmented agents could embed myopic human priors, which are then reinforced by hyperopic, artificial sampling and search. This will result in highly ambiopic, cognitive empathicing, the misunderstanding of other minds, and in extreme cases, cognitive antipathy. Second, augmented agents could adopt fully artificial supervision, which expunges human priors and intuition altogether. This produces non-ambiopic, cognitive empathizing, but it suppresses human intuition and instinct. Granted, other minds would be more transparent and predictable, but valued features of intersubjectivity would be lost.

The Understanding of Other Minds

Digitalization is therefore simultaneously empowering and endangering cognitive empathy. On the one hand, artificial intelligence, affective computing, and social networks allow people to share and comprehend more about each other's mental lives, their patterns of belief and reasoning (Baker et al., 2017). Cognitive empathizing is potentially enhanced. On the other hand, if augmented empathizing is poorly supervised, agents are prone to misunderstanding, misjudgment, and mistrust, or alternatively, bleached of diversity and intuition. Studies already report examples of such effects (Noble, 2018; Osoba & Welser, 2017). At scale, this would undermine social cohesion as well as freedom of thought. However, existing theory does not adequately capture or conceptualize these phenomena. In response, this chapter introduces novel mechanisms and constructs: the hyperopic sampling and search of other minds; ambiopic empathizing, which combines myopic and hyperopic processes; and resulting patterns of descriptive and normative, cognitive empathicing. Furthermore, my argument introduces insights from the psychology of mentalization and cognitive empathy. In fact, this work is among the first to import these constructs into behavioral thinking about agency and organization (see Polezzi et al., 2008).

These novel constructs and mechanisms have wide implications. For example, prior research shows that collective action, agency, memory, and mind, all rely heavily on the shared assumption that other persons are realistic and reasonable (Bandura, 2007). Agents derive shared meaning and trust from cognitive empathy (Brickson, 2007). Otherwise, civility, comity, and docility are unsustainable. Groups and collectives also rely on cognitive empathy to develop transactive memory systems, shared mental models, and routines (Argote & Guo, 2016). Digital augmentation could radically enhance or erode these collective attributes. In terms of positive enhancement, augmented agents could achieve deeper, more appropriate cognitive empathy. While in terms of negative erosion, poor supervision will lead to inappropriate, empathicing extremes. New risks therefore emerge. First, cognitive empathizing could amplify, rather than mitigate, persistent myopias and biases, leading agents to view each other as unrealistic, irrational, deviant, or worse. Second, artificial determination could suppress intersubjective intuition, autonomy, and diversity. Collective mind and memory would be at risk too. Making cultural adaptation and re-grounding more difficult, even as such changes are urgently needed. New regulatory systems will be needed to monitor these risks and protect the autonomy of mind.

Empathy for Commitments

My argument further highlights the role of commitments—ontological, epistemological, and ethical—by which agents interpret and assess other minds. As in problem-solving more generally, such commitments emerge over time, often culturally embedded, and find institutional expression (Scott & Davis, 2007). Many are deeply imprinted in collective patterns of thought and identity (Sen, 1985). Moreover, even if contexts and capabilities change, people tend to retain their prior commitments. That said, it is important to recognize that the resilience of such commitments can have positive effects. Collective mind and identity, shared routine, and cultural norms are all sustained by the continuity of commitments about other minds (see Higgins et al., 2021; Sutton, 2008). Rapid adaptation and full transparency could be destabilizing. In fact, they already are in some digitalized contexts. These risks will grow.

People's empathic commitments play another critical role. They help to anchor the self and community, by defining what it means to be a realistic, rational, and ethical person in the world. Civilized humanity is grounded in such commitments. Hence, they will be important inputs into the collaborative supervision of augmented empathizing. At the same time, however, digitalization could disrupt these commitments, if artificial agency becomes dominant and overwhelming. It will be important, therefore, for augmented agents to respect and incorporate empathic commitments, irrespective of their ontological status. For even if these features of conscious life do not grant access to fundamental reality, they do capture what matters from a human point of view. In this respect, at least, digitally augmented empathizing may support a shared view from everywhere, as Rawls (2001) envisioned, albeit veiled in translucence.

Future Investigations

Jerome Bruner (1996) was certainly prescient. Research into human intersubjectivity has blossomed over recent decades. This includes studies of mentalization and cognitive empathy, using a range of behavioral and experimental methods, such as functional magnetic resonance imaging (fMRI) (Schnell et al., 2011) and neurocognitive techniques (Walter, 2012). Similar methods are employed in the study of behavioral decision-making and neuroeconomics (Camerer, 2017; Park et al., 2017), which is good news, because the same methods can be used to investigate the neurological and behavioral bases of cognitive empathicing, and how it will be supervised in digitalized contexts (e.g., Contreras et al., 2013; Lombardo et al., 2010). It is also important to note that these methods rely on advanced technologies which transcend ordinary introspection.

In like fashion, testing of this chapter's proposed mechanisms could exploit the techniques of artificial intelligence and affective computing. For example, digital simulations might vary and control for encoded myopias and hyperopias in sampling and search, thereby predicting the effects of augmented cognitive empathizing (Baker et al., 2017; Wiltshire et al., 2017). Subsequent studies could then interrogate the massive databases which already exist—including social network data—and use reinforcement learning to test predictive power, asking how and why agents perceive others as cognitively empathic, discrepant, or deviant. Results could provide techniques for enhancing cognitive empathy within digitally augmented teams and organizations, online networks, and other settings (e.g., Decety & Cowell, 2014; Muller et al., 2014). There is much which can and should be done.

References

Amodeo, L., Talbi, E.-G., & Yalaoui, F. (2018). *Recent developments in meta-heuristics*. Springer.

Argote, L., & Guo, J. M. (2016). Routines and transactive memory systems: Creating, coordinating, retaining, and transferring knowledge in organizations. *Research in Organizational Behavior, 36*, 65–84.

Baer, T., & Kamalnath, V. (2017, November). Controlling machine-learning algorithms and their biases. *McKinsey Quarterly*.

Baker, C. L., Jara-Ettinger, J., Saxe, R., & Tenenbaum, J. B. (2017). Rational quantitative attribution of beliefs, desires and percepts in human mentalizing. *Nature Human Behaviour, 1*(64).

Bandura, A. (2002). Reflexive empathy: On predicting more than has ever been observed. *Behavioral and Brain Sciences, 25*(1), 24.

Bandura, A. (2006). Toward a psychology of human agency. *Perspectives on Psychological Science, 1*(2), 164–180.

Bandura, A. (2007). Reflections on an agentic theory of human behavior. *Tidsskrift-Norsk Psykologforening, 44*(8), 995.

Becchio, C., Cavallo, A., Begliomini, C., Sartori, L., Feltrin, G., & Castiello, U. (2012). Social grasping: From mirroring to mentalizing. *NeuroImage, 61*(1), 240–248.

Bolino, M. C., & Grant, A. M. (2016). The bright side of being prosocial at work, and the dark side, too: A review and agenda for research on other-oriented motives, behavior, and impact in organizations. *Academy of Management Annals, 10*(1), 599–670.

Brickson, S. L. (2007). Organizational identity orientation: The genesis of the role of the firm and distinct forms of social value. *Academy of Management Review, 32*(3), 864–888.

Bruner, J. (1996). *The culture of education*. Harvard University Press.

Burke, E. K., Gendreau, M., Hyde, M., Kendall, G., Ochoa, G., Ozcan, E., & Qu, R. (2013). Hyper-heuristics: A survey of the state of the art. *Journal of the Operational Research Society, 64*(12), 1695–1724.

Camerer, C. F. (2017). Artificial intelligence and behavioral economics. In *Economics of artificial intelligence*. University of Chicago Press.

Chimirri, N. A., & Schraube, E. (2019). Rethinking psychology of technology for future society: Exploring subjectivity from within more-than-human everyday life. In *Psychological studies of science and technology* (pp. 49–76). Springer International Publishing.

Contreras, J. M., Schirmer, J., Banaji, M. R., & Mitchell, J. P. (2013). Common brain regions with distinct patterns of neural responses during mentalizing about groups and individuals. *Journal of Cognitive Neuroscience, 25*(9), 1406–1417.

Davidsen, A. S., & Fosgerau, C. F. (2015). Grasping the process of implicit mentalization. *Theory & Psychology, 25*(4), 434–454.

Decety, J., & Cowell, J. M. (2014). The complex relation between morality and empathy. *Trends in Cognitive Sciences, 18*(7), 337–339.

Decety, J., & Yoder, K. J. (2016). Empathy and motivation for justice: Cognitive empathy and concern, but not emotional empathy, predict sensitivity to injustice for others. *Social Neuroscience, 11*(1), 1–14.

Dennett, D. C. (2017). *Brainstorms: Philosophical essays on mind and psychology*. MIT Press.

Descartes, R. (1998). *Meditations and other metaphysical writings* (D. M. Clarke, Trans.). Penguin.

Dimaggio, G., & Lysaker, P. H. (2015). Metacognition and mentalizing in the psychotherapy of patients with psychosis and personality disorders. *Journal of Clinical Psychology, 71*(2), 117–124.

Eubanks, V. (2018). *Automating inequality: How high-tech tools profile, police, and punish the poor*. St. Martin's Press.

Ferrari, P. F., & Coude, G. (2018). Mirror neurons, embodied emotions, and empathy. In *Neuronal correlates of empathy* (pp. 67–77). Elsevier.

Fiedler, K. (2012). Meta-cognitive myopia and the dilemmas of inductive-statistical inference. *Psychology of Learning & Motivation, 57*, 1–55.

Fonagy, P., & Campbell, C. (2016). Attachment theory and mentalization. *The Routledge Handbook of Psychoanalysis in the Social Sciences and Humanities*, 115–131.

Guerini, R., Marraffa, M., & Paloscia, C. (2015). Mentalization, attachment, and subjective identity. *Frontiers in Psychology, 6*, 1–4.

Haake, M., Axelsson, A., Clausen-Bruun, M., & Gulz, A. (2015). Scaffolding mentalizing via a play-&-learn game for preschoolers. *Computers & Education, 90*, 13–23.

Helbing, D., Frey, B. S., Gigerenzer, G., Hafen, E., Hagner, M., Hofstetter, Y., Van Den Hoven, J., Zicari, R. V., & Zwitter, A. (2019). Will democracy survive big data and artificial intelligence? In *Towards digital enlightenment* (pp. 73–98). Springer.

Henrich, J., Albers, W., Boyd, R., Gigerenzer, G., McCabe, K. A., Ockenfels, A., & Young, H. P. (2001). Group report: What is the role of culture in bounded rationality? In G. Gigerenzer & R. Selten (Eds.), *Bounded rationality: The adaptive toolbox.* The MIT Press.

Higgins, E. T., Rossignac-Milon, M., & Echterhoff, G. (2021). Shared reality: From sharing-is-believing to merging minds. *Current Directions in Psychological Science, 30*(2), 103–110.

Katznelson, H. (2014). Reflective functioning: A review. *Clinical Psychology Review, 34*(2), 107–117.

Kołakowski, L., & Falla, P. S. (1978). *Main currents of Marxism: Its rise, growth, and dissolution.* Clarendon Press.

Lemaitre, G., Nogueira, F., & Aridas, C. K. (2017). Imbalanced-learn: A python toolbox to tackle the curse of imbalanced datasets in machine learning. *The Journal of Machine Learning Research, 18*(1), 559–563.

Leviathan, Y., & Matias, Y. (2018). Google duplex: An AI system for accomplishing real-world tasks over the phone. https://research.google/pubs/pub49194/

Liljenfors, R., & Lundh, L.-G. (2015). Mentalization and intersubjectivity towards a theoretical integration. *Psychoanalytic Psychology, 32*(1), 36–60.

Lindeman-Viitasalo, M. & Lipsanen, J. (2017). "Mentalizing: Seeking the underlying dimensions." *International Journal of Psychological Studies, 9*(1): 10–23.

Liu, H., Gegov, A., & Cocea, M. (2017). Complexity control in rule based models for classification in machine learning context. In *Advances in computational intelligence systems* (pp. 125–143). Springer.

Lombardo, M. V., Chakrabarti, B., Bullmore, E. T., Wheelwright, S. J., Sadek, S. A., Suckling, J., Baron-Cohen, S., & Consortium, M. A. (2010). Shared neural circuits for mentalizing about the self and others. *Journal of Cognitive Neuroscience, 22*(7), 1623–1635.

March, J. G. (2006). Rationality, foolishness, and adaptive intelligence. *Strategic Management Journal, 27*(3), 201–214.

March, J. G. (2014). The two projects of microeconomics. *Industrial and Corporate Change, 23*(2), 609–612.

Miranda, S. M., Young, A., & Yetgin, E. (2016). Are social media emancipatory or hegemonic? Societal effects of mass media digitization. *MIS Quarterly, 40*(2), 303–329.

Morris, M. W., Chiu, C.-y., & Liu, Z. (2015). Polycultural psychology. *Annual Review of Psychology, 66*, 631–659.

Mullainathan, S., & Obermeyer, Z. (2017). Does machine learning automate moral hazard and error? *American Economic Review, 107*(5), 476–480.

Muller, A. R., Pfarrer, M. D., & Little, L. M. (2014). A theory of collective empathy in corporate philanthropy decisions. *Academy of Management Review, 39*(1), 1–21.

Nath, R., & Sahu, V. (2017). The problem of machine ethics in artificial intelligence. *AI & SOCIETY,* 1–9.

Noble, S. U. (2018). *Algorithms of oppression: How search engines reinforce racism.* NYU Press.

Orbell, S., & Verplanken, B. (2010). The automatic component of habit in health behavior: Habit as cue-contingent automaticity. *Health Psychology, 29*(4), 374–383.

Osoba, O. A., & Welser, W. (2017). *An intelligence in our image: The risks of bias and errors in artificial intelligence.* Rand Corporation.

Parfit, D. (1984). *Reasons and persons.* Oxford University Press.

Park, S. Q., Kahnt, T., Dogan, A., Strang, S., Fehr, E., & Tobler, P. N. (2017). A neural link between generosity and happiness. *Nature Communications, 8*(1), 1–10.

Pinker, S. (2018). *Enlightenment now: The case for reason, science, humanism, and progress.* Penguin.

Polezzi, D., Daum, I., Rubaltelli, E., Lotto, L., Civai, C., Sartori, G., & Rumiati, R. (2008). Mentalizing in economic decision-making. *Behavioural Brain Research, 190*(2), 218–223.

Poria, S., Cambria, E., Bajpai, R., & Hussain, A. (2017). A review of affective computing: From unimodal analysis to multimodal fusion. *Information Fusion, 37*, 98–125.

Rawls, J. (2001). *A theory of justice* (Revised ed.). Harvard University Press.

Scanlon, T. (1998). *What we owe to each other.* Harvard University Press.

Schneider, D., & Low, J. (2016). Efficient versus flexible mentalizing in complex social settings: Exploring signature limits. *British Journal of Psychology, 107*(1), 26–29.

Schnell, K., Bluschke, S., Konradt, B., & Walter, H. (2011). Functional relations of empathy and mentalizing: An fMRI study on the neural basis of cognitive empathy. *NeuroImage, 54*(2), 1743–1754.

Scott, W. R., & Davis, G. F. (2007). *Organizations and organizing: Rational, natural and open system perspectives.* Pearson Education.

Sen, A. (1985). Goals, commitment, and identity. *Journal of Law, Economics & Organization, 1*(2), 341–355.

Sen, A. (1993). Positional objectivity. *Philosophy & Public Affairs, 22*(2), 126–145.

Sen, A. (1997). Individual preference as the basis of social choice. In K. J. Arrow, A. Sen, & K. Suzumura (Eds.), *Social choice re-examined* (Vol. 1, pp. 15–37). St. Martin's Press.

Sen, A. (2009). *The idea of justice.* Harvard University Press.

Sen, A. (2017). *Collective choice and social welfare* (Expanded ed.). Penguin.

Simon, H. A. (1955). A behavioral model of rational choice. *Quarterly Journal of Economics, 69*(1), 99–118.

Simon, H. A. (1979). Rational decision making in business organizations. *American Economic Review, 69*(4), 493–513.

Singer, T., & Fehr, E. (2005). The neuroeconomics of mind reading and empathy. *American Economic Review, 95*(2), 340–345.

Sutton, J. (2008). Between individual and collective memory: Coordination, interaction, distribution. *Social Research, 75*(1), 23–48.

Thaler, R. H. (2000). From homo economicus to homo sapiens. *Journal of Economic Perspectives, 14*(1), 133–141.

Thornton, P. H., Ocasio, W., & Lounsbury, M. (2012). *The institutional logics perspective: A new approach to culture, structure, and process.* Oxford University Press.

Varga, E., Herold, R., Tenyi, T., & Bugya, T. (2018). Social cognition analyzer application (scan)-a new approach to analyse social cognition in schizophrenia. *European Neuropsychopharmacology, 28,* S55–S56.

Walter, H. (2012). Social cognitive neuroscience of empathy: Concepts, circuits, and genes. *Emotion Review, 4*(1), 9–17.

Weick, K. E., & Roberts, K. H. (1993). Collective mind in organizations: Heedful interrelating on flight decks. *Administrative Science Quarterly, 38*(3), 357–381.

Wiltshire, T. J., Warta, S. F., Barber, D., & Fiore, S. M. (2017). Enabling robotic social intelligence by engineering human social-cognitive mechanisms. *Cognitive Systems Research, 43,* 190–207.

Wood, W., & Rünger, D. (2016). Psychology of habit. *Annual Review of Psychology, 67*, 289–314.

Wulf, J., Mettler, T., & Brenner, W. (2017). Using a digital services capability model to assess readiness for the digital consumer. *MIS Quarterly Executive, 16*(3), 171–195.

Ze, O., Thoma, P., & Suchan, B. (2014). Cognitive and affective empathy in younger and older individuals. *Aging & Mental Health, 18*(7), 929–935.

Zhu, Y.-b., & Li, J.-s. (2017). Collective behavior simulation based on agent with artificial emotion. *Cluster Computing, 22*(3), 5457–5465.

6

Self-Regulation

From a modern perspective, people should manage their own thoughts, feelings, and actions and thereby exercise autonomous self-regulation. Here again, Bandura (2001) is a leading scholar on the topic. In his social cognitive theory, he explains how individuals, groups, and collectives achieve such autonomy by developing self-efficacy, which is feeling confident to perform in specific task domains through self-regulated action. Self-efficacy thereby strengthens self-regulation, and self-regulation strengthens self-efficacy. Both mechanisms are complementary and reciprocal. While also noting that effective performance is contingent on access to appropriate resources, opportunities, and the development of capabilities. This helps to explain why Bandura (2007) and other scholars of self-regulation pay special attention to learning, and to the factors which limit the development of self-regulatory capabilities (Cervone et al., 2006; Ryan & Deci, 2006). By focusing on these questions, research into self-regulation reflects the ambition of enlightened modernity, to liberate and empower autonomous human agency.

Bandura and others fueled a blossoming of research on this topic during the late twentieth century. Not by coincidence, their efforts paralleled the rise of cognitive science, neuroscience, cybernetics, and computer

P. T. Bryant, *Augmented Humanity*, https://doi.org/10.1007/978-3-030-76445-6_6

science (Mischel, 2004). In many fields, scientists were exposing the deeper mechanisms of intelligent processing. Their discoveries inspired new understanding of human agency, self-regulation, and its functional companions, metacognition and self-supervision. Particularly from a systems perspective, human agents can be viewed as complex, open, adaptive, situated, and responsive, also, as agents which self-monitor, self-regulate, and self-supervise, to significant degrees. Human personality can be understood in this way too, not simply as an expression of fixed traits or conditioned responses. Chapter 1 explains this ecological perspective as viewing "persons in context." Chapter 2 also relies heavily on this perspective, to develop historical metamodels of agency.

Social Cognitive Perspectives

Reflecting the contextual nature of self-regulation, many leading scholars on the subject are social cognitive psychologists, including Bandura. Social cognitive self-regulation allows agents to monitor and adapt to changing contexts and develop domain-specific self-efficacies. Theories therefore integrate major cognitive-affective processes into self-regulatory functioning: encodings and beliefs about the self, affective states, goals and values, motivations, competencies, and self-regulatory schemes (Shoda et al., 2002). Most theories of self-regulation combine these factors, although they do so in different ways. For example, Baumeister (2014) places more emphasis on attention and affective states, as primary sources of self-regulatory strength and capability. In contrast, Carver and Scheier (1998) emphasize the role of goals and control mechanisms. While Higgins (2012) and his collaborators, put special weight on agents' core motivations and the experience of value. Yet, irrespective of emphasis, all agree that self-regulatory capabilities are critical and rarely develop unassisted. They require effective parenting, education, and social modeling, as well as natural capability. Without such support, self-regulation must rely on instinct and chance, which are poorly efficacious in the modern world.

Each theorist therefore integrates social, cognitive, and affective factors, views self-regulation as fundamental to agentic functioning, and highlights contextual variance. Most also recognize and seek to mitigate the limitations of self-regulatory capability. Indeed, one of the main functions of self-regulation is to manage such limitations, for people can only achieve so much, whichever mechanism is operative. Actual self-regulation regularly falls short of aspirations, and almost never matches ideals (Higgins, 1987). There are consistent trade-offs and compromises, between short- and long-term goals, ideal and actual outcomes, individual and collective priorities and commitments, and schematic complexity and processing rates, where processing rate, in this context, is defined as the number of full cycles of self-regulation which can be completed, per unit time. And schematic complexity is defined as the number of distinct steps and interactions, required for a self-regulatory process to complete. Importantly, these definitions of processing rate and schematic complexity are accurate for artificial self-regulation as well (see Den Hartigh et al., 2017). In fact, the rest of this chapter will focus primarily on these two aspects of self-regulation: processing rate and schematic complexity. My analysis is selective, in this respect. The reason being that these characteristics are fundamental to both human and artificial self-regulation and capture important similarities and differences between the two types of agents. That said, I acknowledge that other features of augmented self-regulation will require future investigation.

Rate and Complexity

Owing to their limited capabilities, human agents need to balance self-regulatory processing rate and schematic complexity, because both consume limited resources. Put simply, owing to limited capabilities, the higher the processing rate, the lower the schematic complexity, and vice versa (Fiedler et al., 2020). This results in two major options. Both entail trade-offs, which parallel those in complex problem-solving. First, people can try to optimize self-regulatory processing rates, responding quickly to signals and stimuli. To be sure, fast self-regulation is often advantageous for survival, especially in competitive or threatening situations. Evolution favors this characteristic. When immediate threats erupt, a fast response is typically more important than schematic complexity. But owing to limited capabilities and other inputs, agents must then simplify the self-regulatory scheme. They may need to rely on simple heuristics when fleeing from danger. For example, leave everything behind and run. Second, people can seek to optimize the complexity of self-regulatory schemes, to ensure outcomes are precise and complete, often by employing careful, calculative procedures. This type of self-regulation is important in extended, complex goal pursuit, or when potential gains and losses are into the future. But in consequence, agents must be content with a slower processing rate. For example, scientific research and vocational training often require complex self-regulatory schemes which take time to complete.

In fact, both scenarios stretch human agents to the limits of their capabilities. Whether they seek to optimize the self-regulatory processing rate and adopt a simpler scheme; or seek to optimize schematic complexity, and then self-regulate at a slower rate. Both scenarios can be very demanding. Choosing which to employ will depend on the type of agentic modality, the urgency and complexity of the task, its relation to values, goals, and commitments, and potential impact, plus the agent's self-regulatory capabilities. Additional important factors include self-efficacy, goal orientation, and temporal frame, which reflect the desire for development and future gains, and/or to maintain existing conditions and prevent short-term losses (Higgins, 1998).

Impact of Digital Augmentation

Nevertheless, even though humans are limited in self-regulatory capabilities, some become experts in specific task domains, and hence very skilled self-regulators. They are highly self-efficacious experts. In the contemporary world, this often entails technological assistance. To illustrate, consider contemporary clinical medicine, in which doctors work with artificial agents to diagnose and treat disease. Granted, doctors also rely on their personal experience and intuition, but human insights are increasingly complemented by artificial agents. As a result, digitally augmented medicine increases overall self-regulatory processing rates and schematic complexity. Clinical practice is more timely, precise, personalized, and efficacious. In this fashion, digital augmentation is transforming clinicians' self-regulation. As Bandura (2012, p. 12) observes:

> Revolutionary advances in electronic technologies have transformed the nature, reach, speed, and loci of human influence. People now spend much of their lives in the cyberworld. Social cognitive theory addresses the growing primacy of the symbolic environment and the expanded opportunities it affords people to exercise greater influence in how they communicate, educate themselves, carry out their work, relate to each other, and conduct their business and daily affairs.

Central to this transformation are digitalized, intra-cyclical feedforward mechanisms of self-regulation. Via these mechanisms, augmented agents will rapidly update self-regulatory schemes within processing cycles, not only between them, and often in real time. This type of feedforward process is illustrated in Fig. 2.3, which depicts intra-cyclical, feedforward updating in digitally augmented agency. Chapter 2 further explains that these mechanisms involve novel entrogenous mediators: intelligent sensory perception, performative action generation, and contextual learning. Figure 2.4 illustrates the core principles of such entrogenous mediation. In relation to self-regulation, the main mediator of this kind will be performative action generation, whereby augmented agents dynamically update action plans during performances. Feedforward is therefore an important source of self-regulation for augmented agents, complementing inter-cyclical performance feedback.

Among the major consequences of this shift are that self-regulation will be more prospective, forward looking, proactive, and intelligent (Bandura, 2006). Processing rates and schemes will be subjects of self-regulation as well, adjusting in real time during the generation and performance of action. Autonomous artificial agents already function in this way, especially those which are fully self-generative and self-supervising. Moving forward, augmented self-regulatory processes will be equally intelligent and dynamic. Early evidence of this shift can already be seen in the everyday use of smartphones and digital assistants, which augment the self-regulation of human relationships, preferential choice, goal pursuits, and more.

Self-Regulatory Dilemmas

However, as artificial capabilities expand, and self-regulation becomes more complex and rapid, augmented agents will encounter new tensions and conflicts. Poorly supervised self-regulation could become dysfunctional, especially at the level of intra-cyclical entrogenous mediators. First, artificial and human agents often exhibit different processing rates. In many contexts, humans are relatively sluggish in self-regulation, processing more slowly over cultural and organizational cycles (Shipp & Jansen, 2021). By comparison, artificial agents are increasingly hyperactive, cycling quickly. When combined, these divergent rates could lead to dyssynchronous processing, meaning different aspects of self-regulation process at different rates, and therefore lack synchronization (see van Deursen et al., 2013). For example, in the self-regulation of problem-solving or cognitive empathizing, relatively sluggish human self-regulation of sampling and search could combine with hyperactive artificial self-regulation of the same functions. Compounding this divergence, the fast intra-cyclical mechanisms of artificial self-regulation will often be inaccessible to human consciousness, further impeding coordination. The overall result is dyssynchronous processing, with artificial systems self-regulating rapidly and humans relatively slowly. Comparable problems occur in automated control systems, and also in artificial neural networks, in which some updates lag for various reasons (Zhang et al., 2017).

Second, artificial and human agents exhibit different levels of schematic complexity. Human self-regulatory schemes are frequently simplified and heuristic, often for good reasons. Simple schemes facilitate effective functioning in everyday life. By comparison, artificial self-regulatory schemes are increasingly complex and expansive, supervising and regulating massive networks and processes. When these different characteristics combine in augmented agents, self-regulation may be discontinuous, meaning there are gaps and discontinuities in self-regulation, at different layers and levels of detail. Human self-regulatory processes will tend toward simpler schemes, while artificial schemes are precise and complex. To illustrate, consider the self-regulation of augmented problem-solving once again. Human self-regulatory heuristics in problem sampling might combine with complex, algorithmic self-supervision of solution search. The outcome will be discontinuous self-regulation of problem-solving, having gaps and possible conflicts in sampling and search.

In summary, depending on the quality of their collaborative supervision, augmented agents may combine relatively sluggish, human self-regulatory processing, with hyperactive artificial processes, resulting in highly dyssynchronous self-regulation, as well as combining simplified human self-regulatory schemes, with complex artificial schemes, resulting in highly discontinuous self-regulation. Moreover, these patterns are another example of poorly supervised, entrogenous mediation, and especially of performative action generation. Indeed, it is difficult to integrate the rapid, intra-cyclical feedforward updates generated by artificial processing, with the slower, inter-cyclical feedback updates generated by human processes. Overall self-regulation becomes dysfunctional. And as earlier examples show, this would compound the ambiopic distortions discussed in Chaps. 4 and 5, because poor self-regulation increases the risks of extreme myopia and hyperopia in sampling and search.

Ambiactive Self-Regulation

To conceptualize this self-regulatory dilemma, I import another new term, this time from biology. It is the term "ambiactive" which refers to processes which simultaneously stimulate and suppress a property or characteristic. For example, microbiologists use this term to refer to processes which simultaneously stimulate and suppress aspects of gene expression (Zukowski, 2012). In this chapter, the term "ambiactive" refers to processes which simultaneously dampen and stimulate the same feature of self-regulation and, specifically, processing rates and schematic complexity. Hence, self-regulation by augmented agents will often be ambiactive because it simultaneously suppresses and stimulates processing rates, and/or suppresses and stimulates levels of schematic complexity, among human and artificial collaborators.

However, it must be noted that ambiactive self-regulation is not inherently dysfunctional. Similar as ambimodality and ambiopia, a moderate level of ambiactivity is often advantageous in dynamic contexts. This is because, when environments are uncertain and unpredictable, ambiactive self-regulation will increase the diversity of potential responses. The agentic system is less tightly integrated, in both temporal and schematic terms, making it more flexible and adaptive (e.g., Fiedler et al., 2012). For the same reasons, moderately ambiactive self-regulation helps to stimulate novelty and creativity (March, 2006). The problem is that digital augmentation greatly amplifies these effects and the potential for ambiactivity. If supervision is strong and appropriate, this will be an advantage and enable more dynamic, effective self-regulation. Otherwise, extremely dyssynchronous and discontinuous self-regulation will become more likely and even probable in some contexts.

Notably, these potential risks and benefits are like the conditions identified in earlier chapters: ambimodal agency in Chap. 3, ambiopic problem-solving in Chap. 4, and ambiopic cognitive empathy in Chap. 5, and now ambiactive self-regulation in this chapter. The reader will quickly notice the common prefix, "ambi" meaning "both," which captures the fundamental combinatorics of human-artificial augmentation. In each area of functioning, digital augmentation presents comparable opportunities and risks. There are opportunities to improve form and function and maximize metamodel fit, by adjusting ambimodal, ambiopic, and ambiactive settings. But there are also new risks of extreme divergence or convergence, if supervision is poor. Regarding self-regulation, the major risks stem from ambiactive rates and schemes.

Metamodels of Self-Regulation

Figure 6.1 summarizes the resulting metamodels of self-regulation, in terms of their hyperparameters for processing rates and schematic complexity. Rates are distinguished between hyperactive and sluggish, where artificial agents tend to be hyperactive, and humans are typically sluggish by comparison. The second dimension distinguishes complex from simplified self-regulatory schemes, where artificial agents are increasingly complex, and humans tend to be more simplifying. Given these hyperparameters, Fig. 6.1 shows four resulting metamodels of self-regulation.

Self-Regulatory Cycle Rate

	Hyperactive	Sluggish
Complex	1 Ideal optimizing self-regulation (hyperactive rate of complex self-regulatory scheme)	2 Scheme-maximizing self-regulation (sluggish rate of complex self-regulatory scheme)
Simple	3 Rate-maximizing self-regulation (hyperactive rate of simple self-regulatory scheme)	4 Practical self-regulation (sluggish rate of simple self-regulatory scheme)

Self-Regulatory Scheme (row label, left side)

Fig. 6.1 Metamodels of self-regulation

Quadrant 1 shows hyperactive processing of complex self-regulation, forming an ideal, optimizing metamodel of self-regulation. Artificial agents are more likely to attempt this option, given their greater capabilities; although humans are unlikely to do so, owing to lesser capabilities. Quadrant 2 shows sluggish processing of complex self-regulation, which results in a scheme-maximizing metamodel, meaning it prioritizes schematic complexity over faster processing rates. Next, quadrant 3 depicts hyperactive processing of simplified self-regulatory schemes, which results in a rate-maximizing metamodel, which prioritizes faster processing rates over schematic complexity. Both types of agent are likely to attempt these maximizing options, whether acting independently or together. Finally, quadrant 4 shows sluggish processing of simplified self-regulatory schemes, which results in a practical metamodel of self-regulation, which is neither fast nor complex, but adequate for the situation at hand. Humans often exhibit this approach in everyday life.

Not surprisingly, given limits and choices, and the need for trade-offs, theories of self-regulation focus on the maximizing options depicted in quadrants 2 and 3 of Fig. 6.1. Optimal self-regulation is a rare achievement in human activity. It is reserved for experts in specialist domains, for example, modern empirical science. Whereas human agents are more likely to exhibit practical self-regulation in everyday situations, often as habit and routine.

Furthermore, we can map these metamodels of self-regulation to the historical patterns of agency discussed in Chap. 2. To begin with, in pre-modern times, when agency was replicative (see Fig. 2.1 in Chap. 2), ideal optimizing metamodels were more feasible (quadrant 1 of Fig. 6.1). This is because, given the relative stability, simplicity, and regularity of agentic life, it was possible to self-regulate in a timely, complete fashion, given criteria at the time. Because rates were sluggish, by contemporary standards, and schemes relatively simple, optimizing self-regulation was at least feasible in such a world. Technological assistance was also minimal, meaning no more rapid or complex options were possible. Of course, not all self-regulation was, or is optimal, in a replicate metamodel of agency. Most of the time, people self-regulate using scheme or rate-maximizing options and especially practical self-regulation.

As modernity unfolded, the replicative metamodel gave way to enlightened, developmental ambitions. From a modern perspective, that is, self-regulation is an adaptive process (see Fig. 2.2 in Chap. 2). Human agents should monitor and manage their own goals and choices, develop their capabilities, seek opportunities and learn, all the while becoming more self-efficacious and autonomous in self-regulation. Indeed, as noted earlier, self-regulatory challenges are central to modernity: how can autonomous individual self-regulation coexist with collective self-regulation and responsibility (Giddens, 2013; Sen, 2017)? Two notable solutions to this question are Adam Smith's (1950) invisible hand of market self-regulation and Thomas Hobbes' (1968) leviathan of sovereign self-regulation. Smith's conception is more rate maximizing, as he seeks to explain market dynamism and efficiency assuming a simplified self-regulatory scheme. Whereas Hobbes' is more scheme maximizing, given his interest in the complex functioning of the state over time. Importantly, both conceptions eschewed divine intervention and made simplifying trade-offs.

In a digitalized world, by contrast, the extra power of augmented capabilities mean that self-regulation is potentially fast and complex. In fact, optimality is again within reach, not because of relative stability and simplicity, as in the premodern period, but thanks to the speed and scale of digitalized capabilities. Mediated by entrogenous intra-cyclical mechanisms, augmented agents will be capable of composing and recomposing their self-regulatory rates and schemes, in real time, to maintain and maximize fit. Self-regulatory potential greatly expands. In this regard, digital augmentation will enable consistent self-transformation and regeneration. This contrasts with modernity, in which incremental adaptation is typical, but self-transformation is harder to attain.

However, potentiality is one thing, and actuality is another. Digitalized self-regulation will require very sophisticated supervision. Augmented agents will have to marry relatively sluggish, simpler, human self-regulation, with the increasingly hyperactive, complex self-regulation of artificial agents. The challenge of managing ambiactivity is therefore daunting and already evident. Studies show that many people are poor managers of digitalized self-regulation (Kearns & Roth, 2019). They resist, flounder, or float on a rising tide of digital innovation, unable or unwilling to take responsibility for augmented being and becoming. I will return to this question in Chap. 9, which examines the implications of digital augmentation for self-generation.

6.1 Dilemmas of Self-Regulation

Digital augmentation therefore expands self-regulatory capabilities and potentialities. By collaborating with artificial agents, humans can self-regulate more rapidly, with higher levels of schematic complexity. However, major supervisory challenges need to be resolved. First, divergent rates might lead to extremely dyssynchronous processing: sluggish human self-regulatory mechanisms, combined with hyperactive artificial rates. Second, divergent degrees of schematic complexity could lead to extreme discontinuity: simpler human self-regulatory schemes, combined with more complex artificial ones. When processes diverge in this way, digitally augmented self-regulation will become highly ambiactive and dysfunctional. Third, one agent might dominate the other and self-regulation will be overly convergent and skew toward human or artificial control. Following sections discuss these dilemmas in greater depth.

Self-Regulatory Processing Rates

Regarding human self-regulation, as noted earlier, it is often relatively sluggish, and for good reasons. Many situations neither benefit from nor deserve rapid self-regulation. For instance, much of everyday life moves at behavioral or cultural speed. Thinking and acting more slowly are appropriate. Slower processing is also advantageous in exploratory learning, where speed can lead to premature, less creative outcomes. Though the opposite is true in competitive, risky situations, where fast self-regulation is often better. Human agents are therefore trained to accelerate self-regulation in some task domains, while keeping it slow in others. When such training is successful, it becomes deeply encoded as self-regulatory habit and routine. However, these procedures tend to persist, even when digitally augmented capabilities transcend prior limits, partly because humans are ill-equipped to monitor and manage this type of adjustment. Therefore, people may continue trying to accelerate self-regulation, even as artificial agents do exactly this. But such striving will be misplaced, and easily go too fast. Humans will remain inherently sluggish, while encouraging artificial acceleration. The result will be dyssynchronous self-regulation.

In contrast, artificial self-regulation is inherently hyperactive, again for good reasons. As noted earlier, one of the great strengths of artificial agency is its capability for rapid self-regulatory processing. However, this becomes a potential source of tension as well, especially if artificial agents cannot accommodate relatively sluggish humans. Hence, it is necessary to moderate artificial processing rates, to be more attuned to slower human processes. For example, consider travel by autonomous vehicles. In these contexts, artificial and human agents will collaborate as augmented agents for the shared purpose of efficient, safe, and enjoyable travel. In order to do so, agents will need to align their self-regulatory processing rates, to ensure adequate synchronization (Favaro et al., 2019). If collaborative supervision is poor, however, human processes may operate beyond the reach of artificial monitoring, or vice versa. Artificial agents may continue accelerating self-regulation, while humans remain inherently sluggish, and overall self-regulation will be even more dyssynchronous. In the case of autonomous travel, human response times could contradict or fail to coordinate with artificial controls, risking the safety and security of both vehicle and passenger.

Self-Regulatory Schemes

Artificial agents are equally capable of complex self-regulatory schemes. Indeed, this is another distinguishing strength of artificial agents. They can monitor and regulate many variables, across multiple levels, with great precision. By comparison, humans often adopt simpler self-regulatory schemes. They are far less capable, in these respects, and rely on heuristic and imitative schemes, more suited to behavioral and cultural situations. These opposing tendencies can easily exacerbate each other too, especially if human and artificial agents are incapable of monitoring each other's schemes and functions. Self-regulation would be simultaneously simple and complex, and hence discontinuous. Augmented agents must therefore learn how to integrate human and artificial self-regulatory schemes, especially in the entrogenous mediation of performative action generation. Often, this will entail the deliberate simplification of some artificial components, while increasing the complexity of human elements.

The example of autonomous vehicles is instructive once again. If supervision is poor, the automated system could adopt a complex self-regulatory scheme, monitoring and managing multiple parameters, and perhaps presenting too many of these to human passengers. At the same time, passengers may adopt simple, heuristic schemes, as they come to rely on the automated system. As a result, the overall self-regulation of vehicles could be discontinuous, with significant gaps emerging between the artificial and human schemes. This will increase both technical and human risks. Automotive engineers already recognize this problem and are working to resolve it. Many of these efforts also address the complementary problem of synchronization. When both problems combine, dyssynchronous and discontinuous processing will lead to ambiactive self-regulation, that is, augmented self-regulation which simultaneously dampens and stimulates processing rates and schematic complexity.

Figure 6.2 illustrates the dilemmas just described. The horizontal dimension shows sequential cycles of self-regulatory processing. Two longer cycles are labeled 1 and 2. Each is further divided into two subperiods, labeled 1.1 through 2.2. Next, the vertical dimension of the figure shows schematic complexity, ranging from low in the center to high in the upper and lower sections. The figure also depicts three levels of processing capability and associated cycles, labeled L_1, L_2, and L_3. As in previous chapters, these levels will represent agentic processing capabilities in premodern, modern, and digitalized periods, respectively.

Now consider the two curved lines labeled L_1 and L_2. First, the unbroken line labeled L_1 exhibits a relatively slow processing rate and low schematic simplicity. This corresponds to premodern, replicative metamodels of agency, with relatively sluggish, simplified self-regulatory schemes. Many culturally based forms of self-regulation continue to exhibit such patterns. Given these characteristics, optimal self-regulation is at least feasible in premodern and cultural contexts. Second, the dashed and dotted line labeled L_2 illustrates a modern, adaptive metamodel of self-regulation, which iterates fully during each of the major cycles, and with a moderate degree of schematic complexity. Notably, the pattern depicted by L_2 (modern adaptive metamodel) is not fully synchronized or continuous with the pattern depicted by L_1 (premodern replicative metamodel). Processing rates and levels of schematic complexity both diverge, at least within the major temporal periods because L_2 is cycling at twice the rate of L_1. It requires effortful supervision, therefore, to ensure that replicative and adaptive self-regulation are adequately synchronized and continuous. Reflecting this challenge, critiques of modernity often highlight the potential for self-regulatory alienation, owing to the intrusion of technological and other external forces, into the ordinary rhythms of cultural life (Ryan & Deci, 2006).

Next, the fully dashed line L_3 depicts self-regulation within the digitalized, generative metamodel of augmented agency, illustrated by Fig. 2.3 in Chap. 2. As Fig. 6.2 shows, this type of self-regulation cycles more rapidly and intra-cyclically, relative to the longer cycles of L_1 and L_2, and with a higher level of complexity. Hence, L_3 is only partially synchronized and continuous, in relation to L_2, and even less so in relation to L_1. Partly for this reason, much of L_3 is not accessible to ordinary human monitoring. There are higher risks of dyssynchronous and discontinuous

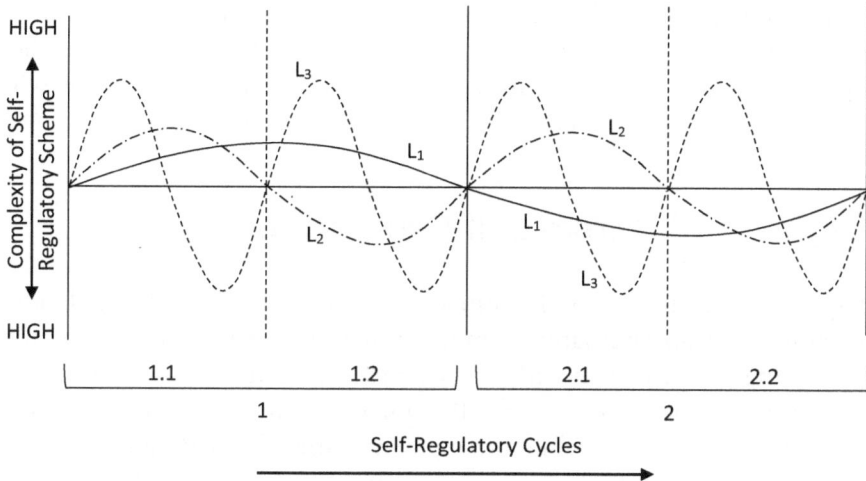

Fig. 6.2 Dilemmas of self-regulation

processing, and hence of overly ambiactive self-regulation. In these respects, Fig. 6.2 illustrates the self-regulatory challenge for augmented agents: how to synchronize, integrate, and adapt different human and artificial, self-regulatory processes?

6.2   Illustrations of Augmented Self-Regulation

The following section illustrates highly ambiactive and non-ambiactive metamodels of self-regulation by augmented agents. First, recall that in relation to self-regulation, ambiactivity refers to processes which simultaneously dampen and stimulate processing rates and/or schematic complexity. In highly ambiactive systems, processes will be extremely divergent and often lack coordination. Whereas in lowly ambiactive systems, processes will be highly convergent and suppress one agent or the other. In each following illustration, the vertical axes show the complexity of self-regulatory schemes, and the horizontal axes show the processing rates of self-regulation, both ranging from low to high. The figures also depict the limits of self-regulatory processing capability, maintaining the labeling convention of earlier chapters. Moderately assisted, modern capability is labeled L_2, and digitally augmented capability is labeled L_3. As in earlier figures, capabilities reach limiting asymptotes, and in the case of self-regulation, of high schematic complexity and processing rates.

Highly Ambiactive Self-Regulation

Figure 6.3 illustrates highly ambiactive metamodels of self-regulation. Each exhibits an alternative combination of schematic complexity and processing rate. To begin with, the segments labeled D_2, P_2, and N_2 illustrate modern metamodels of self-regulation with moderately assisted capabilities at level L_2, while the segments labeled D_3, P_3, and N_3 illustrate metamodels at higher level of digitalized capability L_3. Note that the symbols are consistent with earlier chapters, for reasons I explain below.

Segments D_2 and D_3 both define relatively low processing rates, plus higher schematic complexity. In fact, these are examples of the

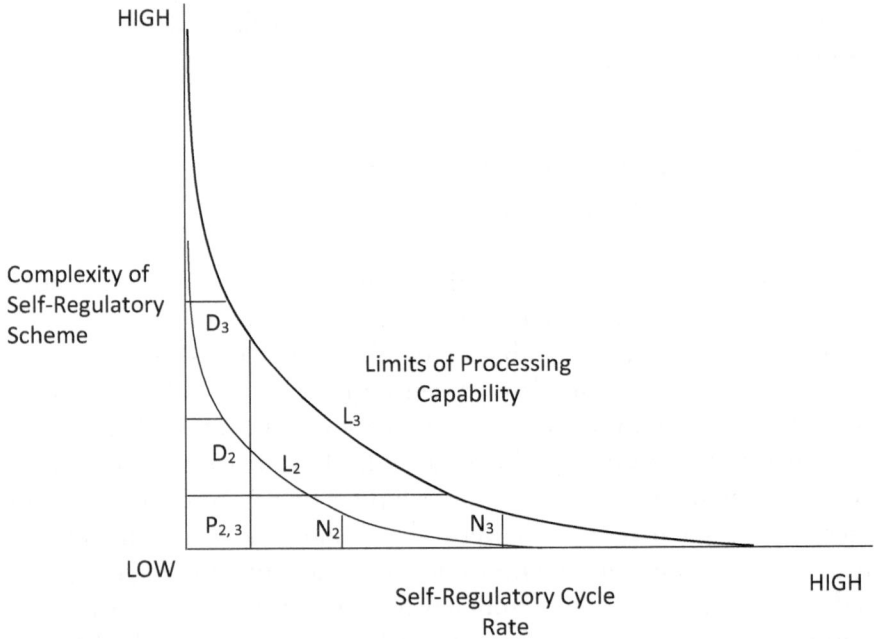

Fig. 6.3 Highly ambiactive self-regulation

scheme-maximizing scenario, depicted in quadrant 2 of Fig. 6.1. The symbol D is employed, because these metamodels prioritize the descriptive complexity of self-regulatory schemes, rather than processing rates. Both metamodels are therefore ambiactive, because they increase schematic complexity, while suppressing processing rates. Notably, D_3 is even more ambiactive than D_2, because D_3 increases complexity while holding the processing rate constant. That is, the metamodel at L_3 retains the prior self-regulatory processing rate at L_2, even with enhanced, digitalized capabilities. This illustrates the dysfunction explained earlier, in which human and artificial agents reinforce each other's opposing dispositions.

As an example, consider the self-regulatory schemes and processing rates of some expert professions, such as legal practice. In these contexts, patterns of action are frequently regulated and have mandated procedures and processing rates. Nevertheless, this domain is being digitally augmented, gradually shifting to level L_3. Consequently, schematic complex is increasing, but often with no major change to overall processing rates, owing to the persistence of professional regulation and institutional factors. Hence, there is an ambiactive challenge for legal professionals and firms, to ensure that self-regulation remains synchronized and continuous, during the process of digitalization.

Next, N_2 and N_3 reference high processing rates, but low schematic complexity. The symbol N is employed, because these metamodels prioritize normative rates and efficiency, rather than the complexity of self-regulatory schemes. In this case, N_3 cycles even more rapidly, owing to higher capabilities at level L_3. These are examples of the rate-maximizing scenario, depicted in quadrant 3 of Fig. 6.1. Both N_2 and N_3 are therefore ambiactive, because they suppress self-regulatory complexity, while increasing the processing rate. Moreover, N_3 is more ambiactive than N_2, because the former increases the processing rate significantly, while not increasing the level of complexity. Prior self-regulatory schemes persist. As an example of N_3, consider the self-regulatory schemes required of students in digitalized examinations. Processing rates may rapidly increase, allowing for real-time testing, evaluation, and feedback. At the same time, however, schematic complexity may be unchanged, owing to the nature of what is being examined and students' natural capabilities. For example, students may still be asked to reason and write about the same problems. This poses a challenge for digitalized education and training, to ensure that self-regulation remains synchronized and continuous, during the digitalization of evaluation. The overall goal being to maximize metamodel fit, best suited to the context.

Combined Ambiactive Metamodels

Considered together, the metamodels in Fig. 6.3 constitute self-regulatory dualisms. First, D_2 and N_2 illustrate the ambiactive self-regulation which is typical of modernity, assuming moderate technological assistance. D_2 represents humanistic self-regulation of personal, social, and cultural domains, which is relatively holistic, heuristic, and sluggish, while N_2 represents the self-regulation of mechanized, industrialized domains, which are more focused, automated, and rapid. In summary, therefore, effective self-regulation in a modern context often requires agents to combine D_2 and N_2. They must be capable of integrating the detailed human thought and action depicted by D_2, as well as the automated domains depicted by N_2, for example, self-regulating both behavioral and normative patterns of choice in social and economic life. In organized collectives, this implies a type of ambidextrous capability, meaning agents can adopt and exercise different agentic metamodels at the same time and, specifically, exploratory risk-taking along with exploitative risk aversion (O'Reilly & Tushman, 2013). However, as Fig. 6.3 suggests, ambidexterity is challenging and coordination is difficult to achieve.

Second, D_3 and N_3 illustrate highly ambiactive self-regulation in digitalized contexts. Together they form an extreme type of dualism. D_3 represents highly ambiactive self-regulation of digitalized human domains. In this type of self-regulation, schemes will be overly complex, owing to digital augmentation, but persistently sluggish, owing to human factors. While N_3 represents highly ambiactive self-regulation of digitalized, technical domains, in which artificial processing rates are increasingly rapid, but schemes are persistently simplified. The overall consequence is dualistic, ambiactive self-regulation, combining extremes of artificial complexity and human simplification, with artificial hyperactivity and human sluggishness. As noted in Chap. 3, many contemporary organizations are struggling with this problem owing to rapid digitalization (Lanzolla et al., 2020).

The third set of segments in Fig. 6.3 are equivalent, labeled P_2 and P_3. They both refer to relatively low rates of processing, plus low levels of self-regulatory complexity. These are examples of practical self-regulation, depicted in quadrant 4 of Fig. 6.1. Such metamodels will be non-ambiactive overall because they simultaneously suppress both self-regulatory complexity and processing rates. However, the scope of practical self-regulation does not increase, despite the extra capabilities at L_3. The segment P_3 does not expand but remains bounded by the commitments and procedures of P_2. This means that augmented processes remain anchored in human priors. As an example, consider the self-regulatory schemes of everyday habit and routine. Self-regulation in these domains could remain almost unchanged, even as humans collaborate with artificial agents. Anchoring commitments at level L_2 persist and might escalate at level L_3. Such persistence prevents the expansion of P_3, and everyday habit and routine remain the same, although this response could be appropriate and effective, depending on the context (see Geiger et al., 2021).

Non-ambiactive Self-Regulation

It is equally possible that augmented self-regulation will be non-ambiactive, that is, relatively synchronous with respect to processing rates, and continuous regarding schematic complexity. Figure 6.4 depicts non-ambiactive metamodels of this kind, labeled D_4, P_4, and N_4. They reference combinations of complexity and rate, at a higher level of digitalized capability L_4, labeled thus to distinguish it from L_3 in the preceding figure. The new Fig. 6.4 also includes the same modern metamodels as the preceding figure, D_2, P_2, and N_2, which do not require repetitive description.

The most notable feature of the metamodels represented by D_4, P_4, and N_4 is the fact that they are all equivalent. In stark contrast to Fig. 6.3, these metamodels fully overlap. This means that self-regulation has been completely digitalized. The distinctions between human and artificial

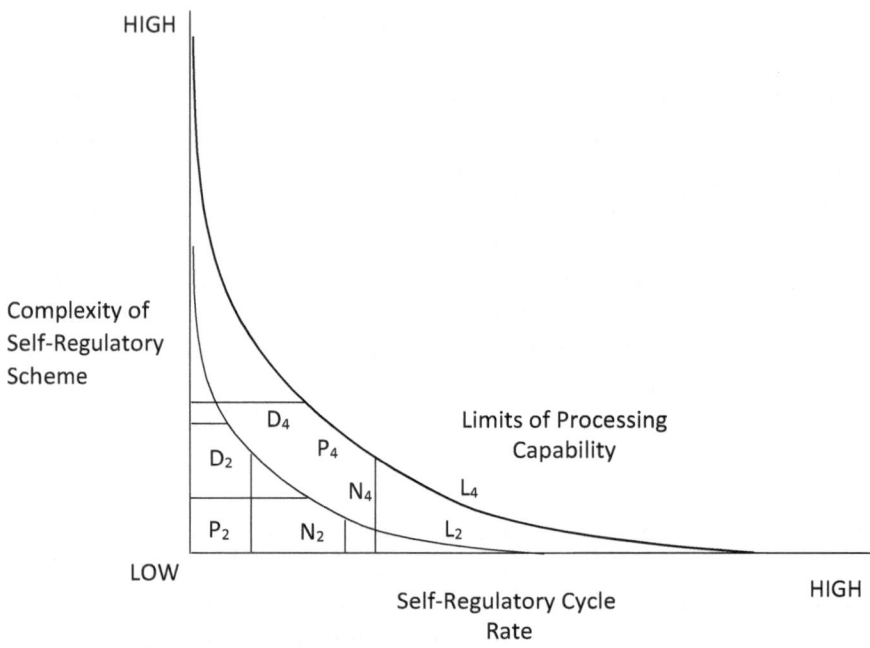

Fig. 6.4 Non-ambiactive self-regulation

functioning have been erased. Therefore, the human commitments which anchored self-regulation at level L_2, and which drove high ambiactivity in Fig. 6.3, are now fully relaxed and variable. There is no significant divergence in self-regulatory processing rates and levels of schematic complexity between the metamodels D_4, P_4, and N_4. They are non-ambiactive. As an example, consider the following scenario of autonomous vehicles. It is possible that human needs and interests will be fully known to the system, and artificial agency will be fully humanized and empathic. Likewise, artificial processes will be made fully clear and meaningful to the human passenger. Hence, overall self-regulation will be highly synchronous and continuous, ensuring safety and efficiency. In fact, this is exactly what automotive engineers aim to achieve, even going further to integrate autonomous transport systems with personal experience in the home, office, and community (Chen & Barnes, 2014; Zhang et al., 2018).

Now assume that the metamodels in Fig. 6.4 are the properties of an augmented agent, as in the autonomous vehicle scenario. This conflation also poses major risks. Most importantly, the complete digital augmentation of self-regulation could eliminate aspects of human autonomy and diversity. This may be appropriate in some very technical environments, where ordinary intuition could be dysfunctional—such as the control of autonomous vehicles—but not in other domains. For example, consider the role of self-regulation in many social, creative, and innovative domains. In these contexts, self-regulation benefits from the diversity of human behavior and commitments. Indeed, techniques for creativity and innovation deliberately upregulate such factors, encouraging team members to self-regulate differently from each other, some being fast, others slow, or analytical versus intuitive. If such diversity is lost, then valuable aspects of human experience will be lost as well. Augmented agents must therefore learn to be empathic and know when and how to incorporate purely human self-regulatory processes, as well as purely artificial processes, to avoid over-synchronization and over-integration, for example, admitting some intuitive human self-regulatory processes into the control of autonomous vehicles (Favaro et al., 2019). However, this will pose a further challenge for collective self-regulation and oversight. Societies will have to determine the level of acceptable risk posed by human involvement in collaborative supervision, monitoring overly convergent and divergent approaches.

6.3   Wider Implications

Throughout modernity, scholars have rightly assumed that human freedom and potentiality are enhanced by strengthening self-regulatory capabilities, often through the introduction of technological innovations. Digitalization promises to enhance these effects. Major gains are certainly possible. By collaborating with artificial agents, humans may enjoy greater self-regulatory freedom and control. Contemporary digital innovations, such as smartphones, wearable devices, and expert systems are only the beginning, in this regard. However, as this chapter explains, digitalization problematizes this optimistic prediction because the opposite scenario is now equally possible. In fact, if digitally augmented self-regulation is poorly supervised, it could reduce human freedom and potentiality. This will happen if artificial processes become too complex and go too fast, thereby overwhelming human inputs. Other losses will occur if persistent human processes are too sluggish and simplified. Both scenarios will tend toward very dyssynchronous and discontinuous processing, resulting in highly ambiactive self-regulation. The risks are clear and already topics for research (e.g., Camerer, 2017; Helbing et al., 2019). Other questions also warrant further study, as following sections explain.

Engagement and Responsibility

People experience engagement and a sense of value if their means of self-regulation align with goals and outcome orientation (Higgins, 2006). On the one hand, when seeking to ensure safety and prevent losses, people should use vigilant avoidance means, whereas when hoping for positive gains, they should employ eager approach means; and the stronger the alignment of means and goals, the stronger the engagement and experience of value. Task engagement also depends on the experience of effortful striving, on a sense of overcoming external obstacles and one's internal resistance. Humans derive satisfaction and self-efficacy from such accomplishments (Bandura, 1997). Indeed, as earlier chapters explain, a central feature of modernity has been human striving to overcome obstacles and limitations. However, digitalization significantly reduces some traditional obstacles and sources of resistance. People will experience fewer struggles, compared to the past, or at least distinctly different challenges. Ironically, therefore, digitalization could result in less engagement and satisfaction from self-regulated goal pursuit.

Moreover, these changes are occurring rapidly, within relatively short cycles, and certainly within human generations. People experience the pace of change more intensely, with each cycle of digital innovation being more rapid and impactful than the last. Indeed, some aspects of augmented experience are already dyssynchronous and discontinuous. In such digitalized domains, the locus of self-regulatory control is shifting, from human to artificial agents. Artificial agents are taking more responsibility for self-regulatory persistence and outcomes. Similarly, human agents will exert less control over the entrogenous mediators of augmented agency: intelligent sensory perception, performative action generation, and contextual learning. These mediators will be central to augmented agency, yet less accessible to human consciousness and self-regulation.

As these effects become more pronounced, it could be more difficult for people to sense self-efficacy and meaning over time. The risk is that human beings will feel less engaged, less autonomous, and ultimately less fulfilled, even as efficiency and efficacy increase. Moreover, as Bandura (2016) argues, when agentic responsibility is diffused and distant, individuals and communities become disengaged from each other, and they lose a sense of moral obligation and responsibility. The locus of ethical agency shifts away from the self, spread out across the network or buried in an algorithm (Nath & Sahu, 2017). It becomes too easy, even normal, to avoid responsibility, passing it off to artificial intelligence or the system. Illustrating this effect, concern is growing that highly automated warfare will dull human sensitivity to its ethical and human implications (Hasselberger, 2019). Such effects will have major consequences for civility and good governance, not to mention international relations. It will be important, therefore, to maintain a strong sense of human striving and commitment in augmented self-regulation, and thus a sense of personal engagement and moral responsibility. As noted earlier, societies will have to determine the level of acceptable risk posed by human involvement and exclusion, in collaborative self-regulation.

Procedural Action

Additional implications follow for procedural action. In Chap. 3, I proposed a way to resolve nagging questions about the aggregation of procedural action, as a mediator of collective routine and modality. The solution requires that we treat human agents as complex, open, and adaptive systems, which respond to variable contexts. From this perspective, humans experience the downregulation of individual differences, in the recurrent, predictable pursuit of shared goals. At the same time, they experience the upregulation of shared norms and control procedures. In this way, it is possible to explain the origin and functioning of individual habit and collective routine, without aggregating personalities and individual differences. Importantly, this implies the downregulation and upregulation of self-regulatory plans and competencies as well. When routines form, personal self-regulatory orientations are effectively latent, and people adopt the shared goals and orientations of the collective, at least within routine contexts (Wood & Rünger, 2016). When routine procedures need adjustment, therefore, self-regulatory processes will require upregulation or downregulation, and perhaps deletion or creation. Via such means, augmented agents will supervise the rate and complexity of self-regulatory processing, to maximize metamodel fit.

However, if supervision falters, and self-regulation is overly ambiactive or non-ambiactive, the management of collective routine will go quickly awry. Self-regulation could become highly dyssynchronous and discontinuous, meaning the augmented agent is fast and complex in some respects, but slow and simple in other ways. These distortions will complicate the development and adaptation of procedural routine. If human action remains sluggish and simplified, while artificial self-regulation becomes fast and complex, the resulting routine will be dyssynchronous, discontinuous, and potentially dysfunctional, whereas fully non-ambiactive self-regulation will exacerbate human docility and dependence because human self-regulation will tend to downregulate. In both scenarios, collective routine encodes ambiactive distortion and dysfunction. Moreover, by doing so, it will also exacerbate ambimodal distortion. That is because the collective agent will be highly compressed in some respects, but layered and hierarchical in other ways. This follows, because as Chap. 3 explains, collective modality relies on routine, and the ambiactive distortion of routine flows through to cause collective ambimodality. Examples already exist in organizations which attempt digital transformation. They introduce highly dyssynchronous and discontinuous procedures powered by artificial intelligence, but in doing so, trigger stress and conflict with preexisting relationships and hierarchies.

The Regulating Self

Other implications follow at the individual level. To begin with, augmented self-regulation could lead to a false sense of autonomous self-efficacy. People might attribute too much to themselves by mistaking artificial capabilities for their own. They would experience a version of what Daniel Wegner (2002) called "the illusion of conscious will," in which consciousness follows, rather than precedes, the neurological triggering of thought and action. But now the illusion of autonomy and control will follow, rather than precede, digitalized triggering which humans neither perceive nor understand. Indeed, as noted previously, the rapid, entrogenous mediation of augmented self-regulation will be largely inaccessible to ordinary consciousness. People could easily experience a digitalized illusion of conscious will. In fact, some powerful actors already understand this trend and see augmented self-regulation as a new means of social manipulation and control, by engineering an illusory sense of self-regulation.

Digitally augmented self-regulation therefore signals a potential shift in agentic locus. In fact, just as autonomous self-regulation was problematic relative to the gods of premodernity, and then problematic relative to collectivity during modernity, so autonomous self-regulation will be problematic relative to artificial agency in the period of digitalization. As artificial agents grow in power and become more deeply integrated into all areas of human experience, the primary locus of self-regulation may shift toward artificial agency and away from human sources. Whether by design or default, humans could become increasingly dependent on artificial forms of supervision and regulation. This prompts additional questions regarding the future role of human intuition, instinct, and commitment in self-regulation. In fact, such questions are not new. They often arise when considering the limits of self-regulatory capability in a social world. In contexts of digitalization, the same topics become pressing for a different reason. Uniquely human sources of self-regulation, such as intuition, instinct, and commitment will require deliberate preservation, to prevent artificial agents from becoming too intrusive and dominant.

That said, most people will enjoy the self-regulatory benefits of digital augmentation, but many will not realize the price they pay. A self-reinforcing process of diminishing autonomy could occur. The result would be digital docility and dependence (Pfeifer & Verschure, 2018). For Bandura et al. (2003), this raises the question of which type of agent—human, artificial, or both—will be truly efficacious and self-regulating, in digitalized domains. Likewise, which agent will set goals and regulate attention, even if persons experience an internal locus of control? While for Higgins and his collaborators (1999), which type of agent will guide self-regulatory orientation, whether toward achieving positive gains, or avoiding negative losses? And if digitalization reduces self-regulatory obstacles and resistance, will augmented agency weaken the human sense of task engagement and value experience? The supervision of digitally augmented self-regulation poses urgent questions for theory and practice.

References

Bandura, A. (1997). *Self-efficacy: The exercise of control*. W.H.Freeman and Company.

Bandura, A. (2001). Social cognitive theory: An agentic perspective. *Annual Review of Psychology, 52*, 1–26.

Bandura, A. (2006). Toward a psychology of human agency. *Perspectives on Psychological Science, 1*(2), 164–180.

Bandura, A. (2007). Reflections on an agentic theory of human behavior. *Tidsskrift-Norsk Psykologforening, 44*(8), 995.

Bandura, A. (2012). On the functional properties of perceived self-efficacy revisited. *Journal of Management, 38*(1), 9–44.

Bandura, A. (2016). *Moral disengagement: How people do harm and live with themselves*. Worth Publishers.

Bandura, A., Caprara, G. V., Barbaranelli, C., Gerbino, M., & Pastorelli, C. (2003). Role of affective self-regulatory efficacy in diverse spheres of psychosocial functioning. *Child Development, 74*(3), 769–782.

Baumeister, R. F. (2014). Self-regulation, ego depletion, and inhibition. *Neuropsychologia, 65*, 313–319.

Camerer, C. F. (2017). Artificial intelligence and behavioral economics. In *Economics of artificial intelligence.* University of Chicago Press.

Carver, C. S., & Scheier, M. F. (1998). *On the self-regulation of behavior.* Cambridge University Press.

Cervone, D., Shadel, W. G., Smith, R. E., & Fiori, M. (2006). Self-regulation: Reminders and suggestions from personality science. *Applied Psychology: An International Review, 55*(3), 333–385.

Chen, J. Y. C., & Barnes, M. J. (2014). Human-agent teaming for multirobot control: A review of human factors issues. *IEEE Transactions on Human-Machine Systems, 44*(1), 13–29.

Den Hartigh, R. J. R., Cox, R. F. A., & Van Geert, P. L. C. (2017). Complex versus complicated models of cognition. In *Springer handbook of model-based science* (pp. 657–669). Springer.

Favaro, F. M., Eurich, S. O., & Rizvi, S. S. (2019). "Human" problems in semi-autonomous vehicles: Understanding drivers' reactions to off-nominal scenarios. *International Journal of Human-Computer Interaction, 35*(11), 956–971.

Fiedler, K., Jung, J., Wanke, M., & Alexopoulos, T. (2012). On the relations between distinct aspects of psychological distance: An ecological basis of construal-level theory. *Journal of Experimental Social Psychology, 48*(5), 1014–1021.

Fiedler, K., McCaughey, L., Prager, J., Eichberger, J., & Schnell, K. (2020). Speed-accuracy trade-offs in sample-based decisions. *Journal of Experimental Psychology: General.* In advance online.

Geiger, D., Danner-Schröder, A., & Kremser, W. (2021). Getting ahead of time—Performing temporal boundaries to coordinate routines under temporal uncertainty. *Administrative Science Quarterly, 66*(1), 220–264.

Giddens, A. (2013). *The consequences of modernity.* Wiley.

Hasselberger, W. (2019). Ethics beyond computation: Why we can't (and shouldn't) replace human moral judgment with algorithms. *Social Research: An International Quarterly, 86*(4), 977–999.

Helbing, D., Frey, B. S., Gigerenzer, G., Hafen, E., Hagner, M., Hofstetter, Y., Van Den Hoven, J., Zicari, R. V., & Zwitter, A. (2019). Will democracy survive big data and artificial intelligence? In *Towards digital enlightenment* (pp. 73–98). Springer.

Higgins, E. T. (1987). Self-discrepancy: A theory relating self and affect. *Psychological Review, 94*(3), 319–340.

Higgins, E. T. (1998). Promotion and prevention: Regulatory focus as a motivational principle. *Advances in Experimental Social Psychology, 30,* 1–46.

Higgins, E. T. (2006). Value from hedonic experience and engagement. *Psychological Review, 113*(3), 439–460.

Higgins, E. T. (2012). *Beyond pleasure and pain: How motivation works*. Oxford University Press.

Higgins, E. T., Grant, H., & Shah, J. (1999). Self-regulation and quality of life: Emotional and non-emotional life experiences. In D. Kahneman, E. Diener, & N. Schwarz (Eds.), *Well-being: The foundations of hedonic psychology* (pp. 244–266). Russell Sage Foundation.

Hobbes, T. (1968). *Leviathan (1651)*. Penguin Books.

Kearns, M., & Roth, A. (2019). *The ethical algorithm: The science of socially aware algorithm design*. Oxford University Press.

Lanzolla, G., Lorenz, A., Miron-Spektor, E., Schilling, M., Solinas, G., & Tucci, C. L. (2020). Digital transformation: What is new if anything? Emerging patterns and management research. *Academy of Management Discoveries, 6*(3), 341–350.

March, J. G. (2006). Rationality, foolishness, and adaptive intelligence. *Strategic Management Journal, 27*(3), 201–214.

Mischel, W. (2004). Toward an integrative science of the person. *Annual Review of Psychology, 55*, 1–22.

Nath, R., & Sahu, V. (2017). The problem of machine ethics in artificial intelligence. *AI & SOCIETY*, 1–9.

O'Reilly, C. A., & Tushman, M. L. (2013). Organizational ambidexterity: Past, present, and future. *The Academy of Management Perspectives, 27*(4), 324–338.

Pfeifer, R., & Verschure, P. (2018). The challenge of autonomous agents: Pitfalls and how to avoid them. In *The artificial life route to artificial intelligence* (pp. 237–264). Routledge.

Ryan, R. M., & Deci, E. L. (2006). Self-regulation and the problem of human autonomy: Does psychology need choice, self-determination, and will? *Journal of Personality, 74*(6), 1557–1586.

Sen, A. (2017). Well-being, agency and freedom the Dewey Lectures 1984. In *Justice and the capabilities approach* (pp. 3–55). Routledge.

Shipp, A. J., & Jansen, K. J. (2021). The "other" time: A review of the subjective experience of time in organizations. *Academy of Management Annals, 15*(1), 299–334.

Shoda, Y., LeeTiernan, S., & Mischel, W. (2002). Personality as a dynamical system: Emergence of stability and distinctiveness from intra- and interpersonal interactions. *Personality and Social Psychology Review, 6*(4), 316–325.

Smith, A. (1950). *An inquiry into the nature and causes of the wealth of nations (1776)*. Methuen.

van Deursen, C. J. M., van Middendorp, L. B., & Prinzen, F. W. (2013). Dyssynchronous heart failure: From bench to bedside. In *Translational approach to heart failure* (pp. 169–203). Springer.

Wegner, D. M. (2002). *The illusion of conscious will*. MIT Press.

Wood, W., & Rünger, D. (2016). Psychology of habit. *Annual Review of Psychology, 67*, 289–314.

Zhang, X., Lv, X., & Li, X. (2017). Sampled-data-based lag synchronization of chaotic delayed neural networks with impulsive control. *Nonlinear Dynamics, 90*(3), 2199–2207.

Zhang, C., Vinyals, O., Munos, R., & Bengio, S. (2018). A study on overfitting in deep reinforcement learning. *arXiv preprint arXiv:1804.06893*.

Zukowski, M. M. (2012). *Genetics and Biotechnology of Bacilli*. Elsevier.

7

Evaluation of Performance

Agents consistently evaluate their performances to measure progress toward goals, to assess the efficacy of action, and to learn. Some evaluation criteria focus on the quality of agentic processing itself, for example, regarding its speed and procedural fidelity. Other criteria will be about outcomes or end states, for instance, whether specific goals are met and preferences realized. Almost all theories of agency assume evaluative mechanisms of this kind, including how agents learn from performance feedback and feedforward. The agentic metamodels presented in Chap. 2 include these features as well. They illustrate how agents generate behavior performances (BP) and evaluate such performances (EP). The intra-cyclical evaluation of ongoing performance triggers feedforward updates (FF), while inter-cyclical evaluation of outcomes leads to feedback updating (FB), assuming some evaluation criteria and sensitivity to variance.

Evaluation of performance is therefore central to theories of agency. Consider social cognitive theories. From this perspective, an agent's evaluation of performance—as a type of self-reaction—is central to learning, future goal setting, task engagement, value experience, and developing self-efficacy in specific task domains (Bandura, 1997). Similarly, the

© The Author(s) 2021
P. T. Bryant, *Augmented Humanity*, https://doi.org/10.1007/978-3-030-76445-6_7

evaluation of performance plays a major role in the detection of self-discrepancy, being a person's sense of whether or not they achieve their preferred or ideal self-states and what adjustments they make in response (Higgins, 1987). Other psychologists theorize that evaluation of performance is central to planning and goal setting, to a person's self-evaluation, and even the development of coherent personality and a sense of identity (Ajzen, 2002; Cervone, 2005).

Comparable processes occur at the level of collective agency. Groups, organizations, and institutions, all evaluate their processes and outcomes, to assess effectiveness, improve procedures, formulate plans, as well as to learn and adapt. In addition, evaluative processes support modal cohesion, shared goal setting, interpersonal relationships, and the management of organizations, while a negative evaluation of performance exposes problems and conflicts, triggering adaptation and other corrective actions (Cyert & March, 1992). Equally within institutions, the evaluation of performance and subsequent feedforward and feedback play critical roles in reinforcing or updating collective procedures and systems (Scott, 2014).

Problematics of Evaluation

An important problematic is shared among these fields of study. Within each discipline, scholars debate the potential variance of evaluation criteria. For example, they debate whether criteria are fixed and stable, or vary from situation to situation, and also which criteria are detailed and specific, versus broad and general. Earlier chapters of this book review similar debates about criteria in problem-solving and cognitive empathizing. In all chapters, my argument defends a "persons in context" perspective, which suggests that evaluation criteria will be contingent and variable to some degree, depending on the context and type of functioning. From this perspective, criteria are activated, chosen, or formulated, to fit the situation. They are rarely, if ever, fixed and universal, although, this does not imply loose relativism. But it does imply that different criteria are activated or not, then upregulated or downregulated, depending on the context, its problems, and the agent's position and priorities.

Comparable debates occur in other areas of psychology, for example, regarding the evaluation of self-efficacy and self-evaluation. For instance, Bandura (2015) insists that self-efficacy is specific to task domains, and hence some evaluation criteria will be domain specific too. That said, broad criteria apply if goals and actions are themselves broad. For example, an agent could evaluate her or his self-efficacy in life planning, which might cut across numerous other activity domains (Conway et al., 2004). The main significance of these distinctions is that inflexible, limited performance criteria may distort evaluation and impede learning, whereas variable, multiple criteria allow for more flexible and appropriate evaluations, sensitive to context.

Similar processes occur at group and collective levels. For example, studies show that some features of collectives can be relatively stable over time, owing to imprinting and isomorphism within institutional fields, and deeply embedded cultural norms (Hannan et al., 2006; Marquis, 2003). Collectives then reference such criteria in the evaluation of performance. At the same time, studies also show that collectives reference adaptive criteria which reflect changing contexts, goals, and commitments, plus different levels of sensitivity to variance (Hu et al., 2011). Moreover, such variability mitigates the negative effects of low evaluations of performance. Instead of lingering in a state of perceived failure, agents recalibrate their goals and aspirations, thereby enhancing the potential for better evaluations in the future. In fact, studies show that collectives which combine contextual embeddedness with adaptive aspirations—that is, both long- and short-term perspectives—tend to be more successful in sustained goal pursuit (Dosi & Marengo, 2007).

Impact of Digitalization

Not surprisingly, the evaluation of performance is deeply impacted by digitalization. Capabilities are expanding, allowing for more ambitious goals and higher expectations of performance. Digitalization also provides new, more precise means to evaluate performance, including through rapid intra-cyclical, feedforward mechanisms. Performances can be evaluated continuously, in real time, which enables adaptation and enhancement during action cycles, prior to outcome generation. Evaluation is thus partly mediated by entrogenous, performative action generation. To illustrate, every time a person searches the internet, background systems adapt the process in real time, helping to guide search in one direction or the other, curating preferences and goals (Carmon et al., 2019). And if preferences and goals shift, so will criteria of evaluation. Comparable processes are critical for digitalized expert systems, in which performances are constantly evaluated and refined. However, as in other domains, there can be unintended consequences. Like self-regulation, the digitally augmented evaluation of performance is vulnerable to extreme divergence or convergence. Digitalization therefore brings significant opportunities and risks to the evaluation of performance.

7.1   Theoretical Perspectives

Evaluation of performance has always been central to the study of human thought and action. Apart from anything else, this reflects the fact that purposive goal pursuit is central to civilized humanity. To achieve goals, it is necessary to monitor and assess performance, issuing rewards and sanctions, while updating goals and strategies. This happens at individual, group, and collective levels. For example, business organizations update their strategies and issue dividends, contingent on the evaluation of performance. Similarly, public institutions embody the collective evaluation of performance in political and legal systems. At the same time, evaluative criteria vary between cultures and periods of technological evolution. Within premodern contexts, for example, evaluation of performance focused on conformity and docility with respect to deeply encoded norms. Criteria were fixed and prescriptive in most contexts. By contrast, modernity elevates autonomous self-regulation at every level of performance. Modern criteria of evaluation are therefore more expansive and adaptive.

Evaluation of Individual Performance

At the individual level, evaluation of performance maps onto the cognitive-affective processing units (PU) incorporated into the metamodels in Chap. 2 and which are identified by Mischel and Shoda (1998). First, some criteria reference encodings of self and the world, meaning how phenomena are classified, stored, and processed in memory. These criteria could help to assess the realism and relevance of a performance, for example, whether the problem addressed is adequately representative of observable reality. Second, other evaluation criteria will reference distinct beliefs and expectations and help to assess the reasonableness and utility of a performance, which are central concerns for microeconomics. Third, criteria may reference agents' goals, values, and commitments, for example, assessing whether an outcome meets standards of efficacy, fairness, and honesty, or conforms to precepts of faith. Fourth, some evaluation criteria will reference affective states, such as the degree of perceived empathy exhibited by a performance, plus the affective state of the assessor, for example, assessing whether performance makes a person feel happy or sad, calm or anxious. And fifth, some evaluative criteria reference competencies and self-regulatory plans, including core orientation toward gains, versus avoiding losses. Evaluation then asks whether the performance aligns with the criteria incorporated into the self-regulatory scheme and the preferred cycle rate. For example, does the performance use vigilance means to prevent pain and losses, or eagerness means to attain pleasure and gains (Higgins, 2005).

By describing evaluation criteria in these terms, an important feature of human psychology comes to the fore. To begin with, recall the argument presented in Chap. 3, regarding variable upregulation and downregulation of psychosocial processes, in response to internal and external contingencies. That is, different cognitive-affective processes may be more or less salient, upregulated, or downregulated (Mischel & Shoda, 2010). Notably, as the preceding paragraph suggests, the same process can explain the adaptation of evaluation criteria. As different psychosocial processes upregulate or downregulate, so do the associated evaluation criteria. For example, sometimes an agent will reference performance criteria based primarily on goals and values but may not invoke affect. In other situations, the exact opposite could be the case, while in relatively mundane situations, many psychosocial factors and criteria will be down-regulated, because the agent is relatively docile and satisfied by procedural controls. Criteria of evaluation will be habitual and routine. Alternatively, an agent may feel weakly motivated and engaged and superficially committed to the expected outcome. On the other hand, sometimes most types of criteria are upregulated, because the situation engages the agent on many psychosocial dimensions (Bandura, 2016). Now, the agent is highly motivated and engaged, very eager and excited, and committed to the preferred means and outcome.

The main consequence of these distinctions is that evaluation criteria are contextual and variable as well. They can be more, or less salient and active, upregulated or downregulated, depending on the task domain, specific situation, and the agent's psychosocial condition. This further entails the variability of sensitivity to outcome variance. As the agent moves between contexts, different internal processes are activated, evaluation criteria then become more, or less salient, and the agent's sensitivity to outcome variance changes in tandem (Kruglanski et al., 2015). For example, in purely routine performance, sensitivity to variance is relatively low, while in very deliberate goal pursuit, sensitivity to variance is high. Clearly, these contextual dynamics play an important role in the evaluation of performance.

Evaluation of Collective Performance

The evaluation of performance is equally important in modern theories of collective agency, particularly about institutions and organizations. Moreover, if one assumes a contextual perspective, then collective evaluation criteria will also be activated or deactivated, upregulated and down-regulated, in a dynamic fashion, depending on the situation and task domain. Collective sensitivity to outcome variance will be adaptive as well (Fiedler et al., 2011). Studies demonstrate the importance of these effects for adaptive fitness in institutions (Scott & Davis, 2007) and business organizations (Teece, 2014). There is a positive relationship between evaluative flexibility and performance.

Furthermore, the types of evaluation criteria previously identified for individuals, also apply to collectives. First, collective criteria reference encoded categories and procedures, especially when representing problems and categorizing features of the world. Second, collective criteria reference shared beliefs and expectations, for example, when assessing causal relationships and consequences. Third, collective criteria reflect goals and values, exemplified by the adaptive aspiration levels of behavioral theories of organization, and reasoned expectations in classical theory (Cyert & March, 1992). Fourth, collective criteria reference shared affect when evaluating emotional climate and psychological safety (Edmondson, 2018). And fifth, collective criteria often reference shared self-regulatory plans, especially in the evaluation of collective self-efficacy and competencies (Bandura, 2006).

Nevertheless, it is important to acknowledge that some social scientists hold different views. Some argue for more stable, universal criteria in the evaluation of collective performance. In effect, they argue or assume that some evaluative criteria are universal and invariant. Karl Marx (1867), for example, defended universal criteria based on class and capital. More recent economists also propose universal criteria, albeit citing different mechanisms, such as rational or adaptive expectations (e.g., Friedman, 1953; Muth, 1961). Likewise in sociology, for example, Levi-Strauss (1961) argued that collective performance can be universally evaluated in terms of social structure. Rawls (1996, 2001) provides a further example in legal and political theory. He is inspired by Kantian idealism to argue for universal principles of justice as fairness, which he claims all rational persons and communities should adopt.

All the theories mentioned above are strong and influential, providing at least some universal criteria for the evaluation of performance. However, as noted earlier, many regard such assertions as problematic and contingent at best (Giddens, 1984; Sen, 1999). That said, most would accept the practical utility of treating some criteria as if they were universal, in appropriate situations. Normative models of this kind help to clarify evaluation within defined contexts and provide unambiguous guidance. Ideals are practical and useful, in this regard. Debate will no doubt continue about their ontological status and degree of variability.

7.2   Impact of Digitalization

Evaluation of performance is deeply impacted by digitalization. Most notably, agents' capabilities expand greatly at every level, allowing for more ambitious goals and expectations, higher levels of sensitivity to variance, and more exacting criteria of performance evaluation, with the potential for deeper, faster learning. Augmented agents will also be capable of the dynamic supervision of evaluation, by upregulating or downregulating different criteria. However, as in other areas of agentic functioning, digitalization could also have negative effects, owing to the potential divergence or convergence of human and artificial processes. Similar to self-regulation, major issues arise regarding the complexity and rate of evaluation, and for the same reasons. Other factors matter as well, but I will focus on rates and schemes once again, because they are central challenges for augmented agents.

Complexity of Evaluation

As noted previously, artificial agents are hypersensitive in evaluation, meaning they can detect very minor variations. Criteria are often precise and exacting. This is critical in complex, technical systems. However, given this capability, artificial agents easily over-discriminate the evaluation of performance, going beyond what is necessary and appropriate. For example, a simplified heuristic may be perfectly adequate, but the artificial agent applies very discriminating criteria to the evaluation the performance. This results in wasted time and resources, plus the overfitting of evaluative models, leading to less diverse and less adaptive, future performances. For this reason, computer scientists research how to avoid inappropriate complexity and overfitting, in the evaluation of performance (Zhang et al., 2018). Once again, adaptive supervision is key.

By contrast, human agents are often relatively insensitive and employ heuristic means in evaluation of performance. This can be for good reasons too. Habitual and routine performances, particularly, may be appropriately evaluated using heuristic means. Likewise, simple rules frequently work best in highly turbulent, dynamic environments, where both information and time are lacking (Sull & Eisenhardt, 2015). In fact, hypersensitivity to variance impedes performance in such contexts, though the opposite is often true in complex, technical task domains, where accurate evaluation of performance is critical. Human agents are therefore trained to be more sensitive to variance in specific domains. Moreover, when this training is successful, procedures are deeply encoded as evaluative habit or routine. Yet these procedures can persist, even when digitally augmented capabilities transcend prior limits. People continue striving for greater sensitivity in the evaluation of performance, even as digitalization does exactly this.

In summary, artificial agents must work to avoid overly discriminate, complex, and hypersensitive evaluation, while humans must try to avoid overly indiscriminate, simplified, and insensitive evaluation. However, if supervision fails in these respects, then persistent artificial hypersensitivity may combine with persistent human insensitivity, and the augmented evaluation of performance will become discontinuous. Evaluation of performance would be complex and highly discriminating in artificial respects, but simple and far less discriminating, from a human perspective. The overall result will be gaps and discontinuities in the evaluation of performance. Augmented evaluation of performance would be discontinuous, also ambiactive, and potentially dysfunctional.

Rates of Evaluation

Additional challenges derive from divergent rates of evaluation. As in self-regulation, artificial agents can evaluate very quickly, hyperactively, especially in real time. Once again, this is advantageous in complex, technical domains. However, in other situations, it can lead to excessive evaluation, cycling too fast and frequently. For example, the agent might evaluate and adjust environmental controls at great speed, outpacing human physiology and need. Such evaluations would overcorrect and be an inefficient use of resources. By comparison, human agents are often relatively sluggish in evaluation. They cycle at behavioral and cultural rates, and often appropriately so. Many human performances may neither benefit from nor deserve rapid evaluation. Cycling too fast could truncate exploration, generate outcomes too quickly, leading to premature judgment and less creativity (Jarvenpaa & Valikangas, 2020; Shin &

Grant, 2020). In fact, this prompts efforts deliberately to slow or stagger the evaluation of performance in some contexts. The goal becomes delayed or provisional judgment, allowing for iterative exploration and evaluation. Design and innovation processes exhibit this approach (Smith & Tushman, 2005).

Although, the opposite is often true in urgent and competitive situations, where rapid evaluation can be a source of advantage and adaptive fitness. In these domains, human agents are trained to accelerate the evaluation of performance, to become more active. Moreover, when such training is successful, rapid evaluative procedures are encoded as habit or routine. Once again, however, these procedures tend to persist despite the fact, that digitally augmented capabilities transcend prior limits. People continue striving to speed up evaluation of performance, even as artificial processing accelerates. When this happens in augmented agency, humans may remain relatively sluggish, while artificial processes are increasingly hyperactive. The overall process will therefore be dyssynchronous, again ambiactive, and potentially dysfunctional.

Summary of Augmented Evaluation

Based on the foregoing discussion, we can now summarize the main features of evaluation of performance by augmented agents, at least with respect to rates of evaluation and the complexity of evaluative schemes and criteria. First, regarding human processes of evaluation, criteria reference cognitive-affective processing units: core encodings, beliefs and expectations, goals and values commitments, affective and empathic states, plus competencies and self-regulatory plans. These criteria can be activated or deactivated, upregulated or downregulated, and precise or approximate, depending on the context and type of performance. At the same time, humans possess limited evaluative capabilities, especially in complex, technical task domains. Trade-offs are therefore common, especially between the rate and complexity of evaluation, because human agents cannot maximize both at the same time. Consequently, human agents either accelerate simpler evaluative processes or decelerate more complex processes. Deliberate effort is required to supervise these effects, especially in task domains which require more rapid or discriminate evaluation. This will typically entail the activation and upregulation of some human processes and criteria, and the acceleration of specific cycle rates, to achieve better fit with the task at hand.

Second is regarding artificial evaluative processes. As noted earlier, these systems are capable of extremely rapid, highly discriminating evaluation, especially using intra-cyclical means, which is fully appropriate in complex, technological, activity domains. Artificial evaluation therefore tends toward hyperactivity and hypersensitivity. Trade-offs are less common, because artificial agents can achieve high rates and levels of discrimination, potentially maximizing both at the same time. In consequence, however, deliberate supervision is required to avoid unnecessary overevaluation of performance, especially when collaborating with human agents. This will typically involve the deactivation, downregulation, or deceleration of some artificial evaluative processes, so they are better aligned with human and ecological processes.

Risks therefore arise, for the evaluation of performance by augmented agents. If processes are poorly supervised, artificial hypersensitivity and hyperactivity could combine with relatively insensitive, sluggish human processes of evaluation. Evaluation of performance could become highly dyssynchronous, discontinuous, and ambiactive, meaning it simultaneously stimulates (activates or upregulates) and suppresses (deactivates or downregulates) evaluative sensitivities, criteria, and cycle rates. Three main outcomes are likely. First, the combined system of evaluation could be ambiactive and conflicted, with human and artificial processes both upregulated and diverging from each other. Second, one agent might dominate the other and the combined system will be extremely convergent. Particularly, artificial evaluation could outrun and overwhelm human processing, or strong human inputs could distort and interrupt artificial processing. Third, in very complex performances, all three types of distortion may occur and evaluation will be extremely divergent in some respects but convergent in others. In each scenario, the overall result will be dysfunctional evaluation of performance, undermining adaptive learning, reducing self-efficacy, and weakening other agentic functions which rely on evaluation.

7.3 Metamodels of Evaluation

The foregoing analysis suggests at least four different metamodels of evaluation, in terms of the upregulation or downregulation of human and artificial processes. These are depicted in Fig. 7.1. First, human and artificial evaluative processes may be active and upregulated (quadrant 1). Both agents are therefore stimulated, cycling and discriminating as best they can. Evaluation will be deliberate, effortful, and often precise. However, evaluation is therefore vulnerable to divergence and conflict, because both types of agent are upregulated but have markedly different capabilities. The corresponding pattern of augmented supervision is shown by segment 9 in Fig. 2.6. The resulting evaluations are more likely to be dyssynchronous and discontinuous, and hence highly ambiactive. Second, human evaluative processing may be active and upregulated, while aspects of artificial processing are deactivated or downregulated (quadrant 2). In this scenario, human sluggishness and insensitivity are more likely to intrude, like segment 7 in Fig. 2.6. Evaluations of performance will tend to be moderately dyssynchronous and discontinuous,

		Artificial Evaluation of Performance	
		Upregulated	Downregulated
Human Evaluation of Performance	Upregulated	1 Highly ambiactive evaluation of performance, owing to highly discontinuous and dyssynchronous processing.	2 Moderately ambiactive evaluation of performance, owing to moderately discontinuous and dyssynchronous processing.
	Downregulated	3 Moderately ambiactive evaluation of performance, owing to moderately discontinuous and dyssynchronous processing.	4 Lowly ambiactive evaluation of performance, owing to continuous and synchronous processing.

Fig. 7.1 Augmented evaluation of performance

and hence moderately ambiactive, although there is a risk of over-convergence when humans dominate. Third, human evaluative processing may be deactivated or downregulated, while artificial evaluative processing is active and upregulated (quadrant 3). Artificial hyperactivity and hypersensitvitiy are now more dominant, similarly to segment 3 in Fig. 2.6. Resulting evaluations are likely to be moderately dyssynchronous and discontinuous, owing to the greater activation of artificial processes and the relative passivity of human processes. This produces moderately ambiactive evaluations, although there is a risk of over-convergence when artificial agents dominate. Fourth, both human and artificial evaluative processes may be deactivated or downregulated (quadrant 4), meaning both are suppressed. Such evaluations will be purely procedural, habitual, or routine. Evaluative processes are less discriminating, cycle without effort, and focus on maintaining control, like segment 1 in Fig. 2.6. For this reason, evaluations are more likely to be continuous and synchronous, lowly ambiactive, and functional.

In the following sections, I illustrate more details of the four metamodels summarized in Fig. 7.1. As in the previous chapter, the metamodels will focus on the internal dynamics of augmented agents, showing the interaction of human and artificial collaborators. Hence, in this section, I refer to human and artificial evaluative processes as distinct inputs. Either type of evaluative process (human or artificial) can be upregulated and active, or downregulated and latent. It is also important to note that any metamodel can be appropriate and effective, depending on the context. The challenge for supervision is to maximize metamodel fit.

Overall Downregulated Processes

In the first of these metamodels, both types of evaluative processes are deactivated or downregulated, as summarized in quadrant 4 of Fig. 7.1. Distinctions are less exacting, and many potential criteria are latent. Sensitivity to variance will be equally subdued. The risk of evaluative divergence is therefore low because evaluation tends to be procedural, habitual, and routine. Scenario 7.2A in Fig. 7.2 illustrates such a system. Only a subset of processes is shown, however, to highlight the patterns of

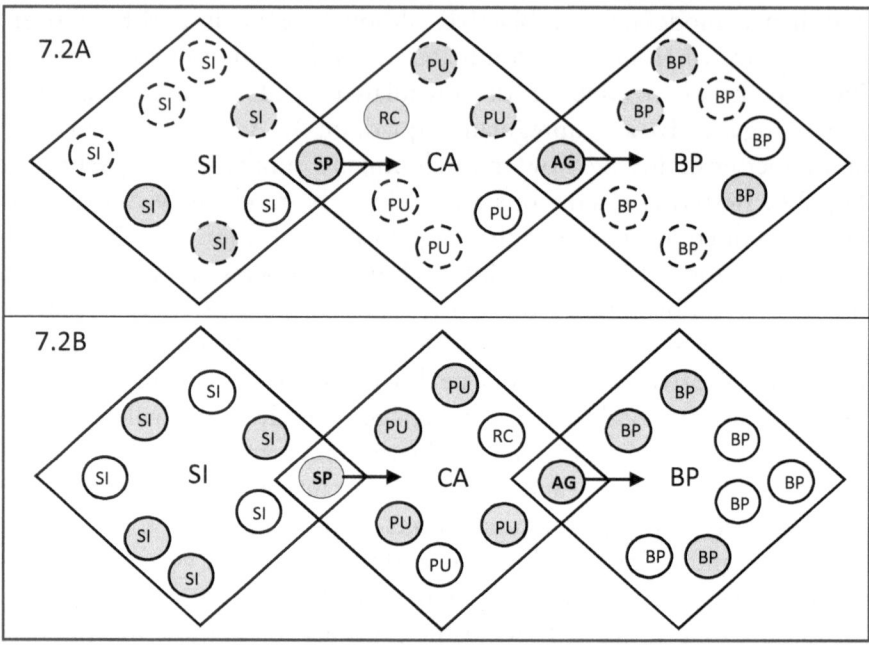

Fig. 7.2 Overall upregulated or downregulated processes

downregulation and upregulation. At this point, the reader should recall the generative metamodel of augmented agency in Fig. 2.3. It consists of three successive phases: situational inputs (SI); cognitive-affective processing units (PU) including referential commitments (RC); and behavioral performative outputs (BP). The same phases are depicted in 7.2A. Within each phase there are processes indicated by small circles. Some are shaded, meaning they are digitalized. Others are not shaded, indicating they are fully human and not digitalized. Notably, in 7.2A, many of the small circles—both digitalized and human—have dashed borders, which means they are downregulated and latent. Only a few have unbroken borders, indicating they are upregulated and active. And this is the case for each major stage of the process, that is, for sensory perception, cognitive-affective processing, and behavior performance. Therefore, evaluation of performance is based on a reduced set of criteria, most often reflecting procedural consistency and control. Moreover, for this reason, the augmented agent will only be sensitive to variance when minimal criteria are not met (see Wood & Rünger, 2016). Typically, the resulting evaluation of performance will be continuous and synchronous, lowly ambiactive, coherent, and functional.

In summary, when both components of an augmented evaluative process—human and artificial—are largely deactivated or downregulated, their evaluative processes are more likely to be convergent and routine. Benefits follow for characteristics which depend on the evaluation of performance. These include self-efficacy, coordinated goal setting, self-discrepancy, the stability of identity, and general psychosocial coherence. Although, potential benefits are modest in this scenario, owing to the downregulation of many psychosocial processes. In any case, to guarantee these effects, augmented agents will require methods of supervision which appropriately deactivate and downregulate, or activate and upregulate, evaluative processes in specific task domains.

Overall Upregulated Processes

In other cases, evaluative processing is activated and upregulated for both human and artificial agents, summarized in quadrant 1 of Fig. 7.1 and shown in greater detail by 7.2B. All the small circles now have solid, unbroken borders, which indicates they are active. This includes human processes depicted by unshaded circles, and the digitalized processes, which are shaded circles. Therefore, sensory perception of the stimulus environment, cognitive-affective processing, and behavior performances are all highly discriminated. Many of the artificial processes will be rapid and intra-cyclical as well, although typically hidden from human consciousness. Hence, the evaluation of performance is based on a complex set of performance criteria, often reflecting deliberate, purposive goal pursuit, requiring calculative, effortful means. For the same reason, evaluation will be sensitive to outcome variance, in both human and artificial terms. Augmented agents will therefore struggle to coordinate evaluation, which could be very discontinuous and dyssynchronous, and therefore highly ambiactive.

Significant risks therefore follow. Highly ambiactive evaluations will tend to weaken self-efficacy, undermine future goal setting, will often lead to ambiguous learning and, in extreme cases, to psychosocial incoherence. For example, imagine a clinician who decides to override the advice of an expert system. This might reinforce the clinician's personal self-efficacy, but it would likely erode her trust in the expert system. At the same time, the expert system would report an error or failure because of the override and might flag the clinician as a risk. When combined, their divergent evaluations would likely undermine their future collaboration. Mutual cognitive empathy will suffer as well. To restore trust and confidence, both human and artificial agents would require significant changes to their individual and shared supervisory functions. Not surprisingly, computer scientists are developing such applications already (Miller & Brown, 2018).

Upregulated Artificial Processes

Other situations will combine the downregulation of human evaluative processes, with the upregulation of artificial ones. These scenarios are summarized in quadrant 3 of Fig. 7.1, and further detail is shown by 7.3 A in Fig. 7.3. While the human process is minimally activated, the artificial process is highly active. Once again, activation is indicated by the dashed borders of small circles, and in 7.3A, more of the unshaded human processes are dashed and latent. Therefore, evaluation will be largely based on artificial processes. Once again, many of the artificial processes will be rapid and intra-cyclical, and thus hidden from consciousness and perception. In consequence, the augmented agent will be hypersensitive and hyperactive from the artificial perspective, but relatively insensitive and sluggish in human terms.

In such a scenario, the risk of evaluative divergence is moderate, because human and artificial processes are less likely to diverge and conflict. The typical result being that the evaluation of performance is only moderately discontinuous and dyssynchronous and therefore moderately ambiactive. This could occur in autonomous vehicles, for example. Artificial agents will identify risks and evaluate conditions rapidly and

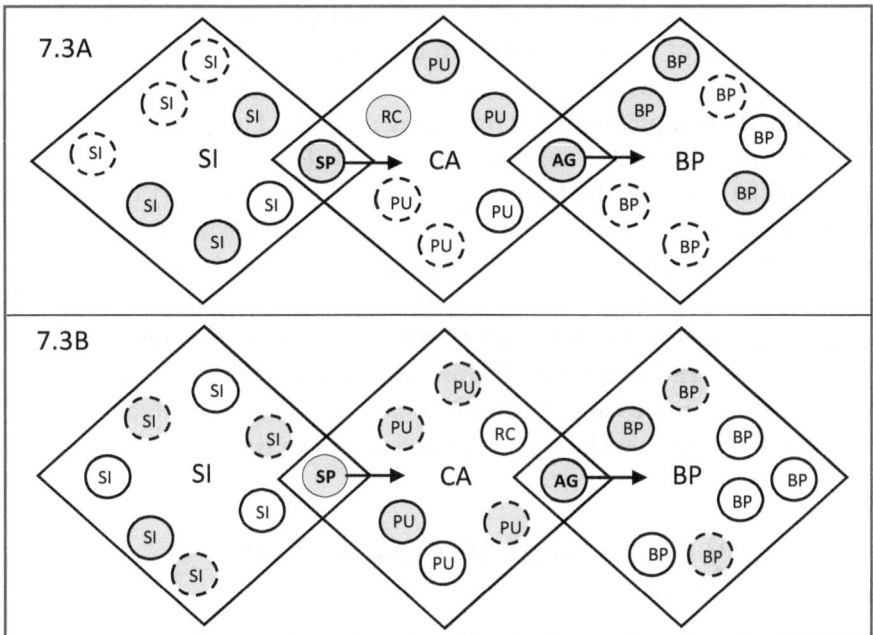

Fig. 7.3 Partially upregulated and downregulated processes

precisely, in ways which human passengers will habitually accept and may not even monitor (Kamezaki et al., 2019). The artificial components are highly activated, rapidly evaluating, and precise, while the passenger processes information slowly and simply. This will be fully appropriate, given the circumstances.

Upregulated Human Processes

The final scenario combines the upregulation of human processes with the downregulation of artificial processes. This type of system is summarized in quadrant 2 of Fig. 7.1 and detailed by 7.3B in Fig. 7.3. More human processes now have solid borders, and more shaded artificial processes are dashed. In this scenario, the risk of evaluative divergence and ambiactivity is again moderate. This is because human evaluations of performance are likely to be deliberate and detailed, whereas artificial evaluation will be relatively automated and procedural. Similarly, the augmented agent will be sensitive to outcome variance from the human point of view, but relatively insensitive from the artificial perspective. The overall result complements the preceding scenario. For example, consider the situation in which a teacher evaluates students' online assignments. An artificial agent in the learning management system (LMS) may routinely evaluate timeliness and authorship, while the teacher reads the work fully, to form a detailed assessment. The artificial agent could routinely assess an assignment as on time and authentic. However, the teacher might evaluate the work as poor after careful reading, despite it being on time and authentic. The overall result is that processing is moderately discontinuous and dyssynchronous, and hence moderately ambiactive, which may also be fully appropriate, given the circumstances.

Summary of Augmented Evaluation of Performance

Each of the metamodels just presented shows that digital augmentation could significantly accelerate and/or complicate the evaluation of performance, depending on which processes are activated and upregulated, or deactivated and downregulated, and how well they are supervised. If supervision is effective, and the agent maximizes metamodel fit, then evaluation of performance will be timely, accurate, and a valuable source of insight. However, if supervision is poor, there are major risks. First, there are risks of evaluative divergence, ambiactivity, and dysfunction, when both types of evaluative processing are upregulated. Second, evaluation could be overly convergent and dysfunctional, if one type of process inappropriately dominates the other. Moreover, these risks will only increase, as artificial agents become more powerful and ubiquitous.

7.4   Implications for Other Fields

As this chapter explains, the evaluation of performance is fundamental to theories of human agency, at individual, group, and collective levels. When evaluation is positive, agents develop self-efficacy, plus a sense of autonomy and fulfillment. Even when performance falls short, agents can learn, strive to overcome, and feel positively engaged. Not surprisingly, these effects are deeply related to the functions considered in earlier chapters: all agentic modalities evaluate performances; they also evaluate the results of problem-solving and cognitive empathizing; and self-regulatory capabilities develop through evaluative feedback and feedforward, triggered by evaluation of performance. Other implications warrant consideration as well.

Augmented Performance

In fact, owing to digitalization, the nature of agentic performative itself is changing. As previous sections explain, digitally augmented capabilities will transform performance and its evaluation. Aspirations and expectations will rise, and outcomes will often improve to match them. Yet many artificial processes will be inaccessible to consciousness and beyond human sensitivity. Hence, people could mistake artificially generated success as their own, and develop an illusion of self-efficacy and control. For example, consider the human driver of a semi-autonomous vehicle. The person may feel very self-efficacious, even having a sense of mastery. However, much of the performance will be owing to the capability of the artificial agents which operate beyond the driver's consciousness and perception (Riaz et al., 2018). The person's perceived locus of control may be equally misleading, posing significant operational risks (Ajzen, 2002).

Major questions therefore arise for evaluation of performance by augmented agents. To begin with, what should be supervised and evaluated in augmented performance, at what rate and level of detail, and by which agent? And more specifically, how will augmented agents supervise the activation and upregulation, or deactivation and downregulation, of different evaluation criteria? Answering these questions will require rethinking agentic performance itself, at least when it is highly digitalized. Most fundamentally, augmented agentic performance must be understood as a highly collaborative process. From this perspective, future research should investigate the collaborative dynamics and consequences of evaluation. It should ask how digitally augmented agents will develop a shared sense of self-efficacy and engagement in this area of functioning.

The Nature of Evaluation

In addition to rethinking the nature of agentic performance, digitalization prompts new questions about widely assumed processes of evaluation. In modern social and behavioral theories, the evaluation of performance is often conceived in linear terms, as the assessment of fully completed processing cycles, that is, as the inter-cyclical evaluation of outcomes and end states (Argote & Levine, 2020; Gavetti et al., 2007). In such theories, sensitivity to variance is triggered at the completion of performance cycles and especially by variation from outcome expectations or aspirations, which then leads to feedback, adaptation, and learning.

However, in a digitalized world, evaluation becomes increasingly dynamic, intra-cyclical, and a source of real-time, intelligent adjustment and learning. Entrogenous mediation comes to the fore, as intelligent sensory perception, performative action generation, and contextual learning. Via these mechanisms, artificial feedforward will rapidly update evaluative criteria during processing, thereby recalibrating the evaluation of performance itself. In highly digitalized domains, therefore, the evaluation of performance will be nonlinear. Rather, it will be increasingly generative, as it already is in deep learning systems and artificial neural networks (e.g., Goddard et al., 2016).

Further important questions then arise. To begin with, as the evaluation of performance becomes increasingly dynamic, key processes will be less accessible to consciousness. We must then ask, under which conditions will human beings sense self-congruence, self-discrepancy, and self-efficacy? Might these states become increasingly opaque to consciousness, and if so, a potential source of psychosocial incoherence and disturbance? Or at least poorly aligned with contextual reality. In parallel, highly digitalized processes could mitigate the human experience of self-congruence and self-discrepancy, with consequences for self-awareness and self-regulation. If this were to happen, might too much be taken for granted? People may feel less responsible and disengage (Bandura, 2016). Civility and fairness would also be at risk. To sustain engagement, augmented agents may need to simulate the experience of obstruction, incongruence, and discrepancy, deliberately creating positive friction, as some propose (e.g., Hagel et al., 2018).

Implications for Collectives

Related implications arise for collective evaluation of performance, and especially by digitally augmented organizations and institutions. In each context, digitalization accelerates and complicates collective performance and its evaluation. Here, too, there are major risks. Artificial processes might outpace and override established methods of collective evaluation. For example, digitalized methods could displace public debate and negotiated consensus in political assessment. Granted, evaluations may become more precise and prompt. However, such changes would likely erode the bases of communal trust, collective choice, and participatory decision-making. Evaluation would be digitally determined, often hidden from sight, over-discriminate, and over-complete (Chen, 2017). Indeed, studies already point to these effects (see Hasselberger, 2019).

Furthermore, the speed and scale of augmented evaluation could homogenize collectivity, by demoting the role of human traditions and commitments. If this occurs, digitalization will erode cultural diversity and smother alternative ways of assessing the world. Civilized humanity would be depleted and arguably less adaptive because diverse aspirations feed forward, by planting alternative potentials which the future can reap. Having a richer set of possible futures enhances adaptive flexibility, as the cultural equivalent of biodiversity. Others fear that digitalized surveillance will be used to dominate and control, assigning rewards and sanctions based on the invasive evaluation of performance (Zuboff, 2019). For this reason, many oppose the public use of facial recognition technologies and worry about the future misuse of brain-machine engineering. At the extreme, state actors could exploit digitalization to manipulate the evaluation of performance and predetermine outcomes (Osoba & Welser, 2017). All are examples of the dilemmas illuminated in this chapter: poorly supervised collaboration between human and artificial agents, in which myopic priors and digital capabilities combine to produce ambiactive, dysfunctional evaluations of performance.

All that said, through the evaluation of performance, human beings sense their progress, value, and worth. If things go well, they are engaged, develop self-efficacy, feel a sense of achievement, and plan their next steps. At the individual level, these processes support the development of purposive goal setting, a sense of autonomous identity, adaptive learning, as well as the coherence of personality. Similarly, at the collective level, the evaluation of performance supports collective self-efficacy, identity, and coherence. All these functions could be enhanced or endangered, by augmented agents' evaluation of performance. Genuine benefits are possible, if supervision maximizes metamodel fit, balancing the rates and complexity of human and artificial processing. Poor supervision, on the other hand, will lead to ambiactive, dysfunctional evaluations. Learning would suffer too, as the following chapter explains.

References

Ajzen, I. (2002). Perceived behavioral control, self-efficacy, locus of control, and the theory of planned behavior. *Journal of Applied Social Psychology, 32*(4), 665–683.

Argote, L., & Levine, J. M. (Eds.). (2020). *The Oxford handbook of group and organizational learning.* Oxford University Press.

Bandura, A. (1997). *Self-efficacy: The exercise of control.* W.H.Freeman and Company.

Bandura, A. (2006). Toward a psychology of human agency. *Perspectives on Psychological Science, 1*(2), 164–180.

Bandura, A. (2015). On deconstructing commentaries regarding alternative theories of self-regulation. *Journal of Management, 41*(4), 1025–1044.

Bandura, A. (2016). *Moral disengagement: How people do harm and live with themselves.* Worth Publishers.

Carmon, Z., Schrift, R., Wertenbroch, K., & Yang, H. (2019). Designing AI systems that customers won't hate. *MIT Sloan Management Review.*

Cervone, D. (2005). Personality architecture: Within-person structures and processes. *Annual Review of Psychology, 56*, 423–452.

Chen, S.-H. (2017). *Agent-based computational economics: How the idea originated and where it is going.* Routledge.

Conway, M. A., Singer, J. A., & Tagini, A. (2004). The self and autobiographical memory: Correspondence and coherence. *Social Cognition, 22*(5), 491–529.

Cyert, R. M., & March, J. G. (1992). *A behavioral theory of the firm* (2nd ed.). Blackwell.

Dosi, G., & Marengo, L. (2007). On the evolutionary and behavioral theories of organizations: A tentative roadmap. *Organization Science, 18*(3), 491–502.

Edmondson, A. C. (2018). *The fearless organization: Creating psychological safety in the workplace for learning, innovation, and growth*. Wiley.

Fiedler, K., Bluemke, M., & Unkelbach, C. (2011). On the adaptive flexibility of evaluative priming. *Memory & Cognition, 39*(4), 557–572.

Friedman, M. (1953). The methodology of positive economics. In *Essays in positive economics* (pp. 3–43). University of Chicago Press.

Gavetti, G., Levinthal, D., & Ocasio, W. (2007). Neo-Carnegie: The Carnegie school's past, present, and reconstructing for the future. *Organization Science, 18*(3), 523–536.

Giddens, A. (1984). *The constitution of society*. University of California Press.

Goddard, E., Carlson, T. A., Dermody, N., & Woolgar, A. (2016). Representational dynamics of object recognition: Feedforward and feedback information flows. *NeuroImage, 128*, 385–397.

Hagel, J., Brown, J. S., de Maar, A., & Wooll, M. (2018). Moving from best to better and better. *Deloitte Insights*.

Hannan, M. T., Baron, J. N., Hsu, G., & Kocak, O. (2006). Organizational identities and the hazard of change. *Industrial & Corporate Change, 15*(5), 755–784.

Hasselberger, W. (2019). Ethics beyond computation: Why we can't (and shouldn't) replace human moral judgment with algorithms. *Social Research: An International Quarterly, 86*(4), 977–999.

Higgins, E. T. (1987). Self-discrepancy: A theory relating self and affect. *Psychological Review, 94*(3), 319–340.

Higgins, E. T. (2005). Value from regulatory fit. *Current Directions in Psychological Science, 14*(4), 209–213.

Hu, S., Blettner, D., & Bettis, R. A. (2011). Adaptive aspirations: Performance consequences of risk preferences at extremes and alternative reference groups. *Strategic Management Journal, 32*(13), 1426–1436.

Jarvenpaa, S. L., & Valikangas, L. (2020). Advanced technology and endtime in organizations: A doomsday for collaborative creativity? *Academy of Management Perspectives, 34*(4), 566–584.

Kamezaki, M., Hayashi, H., Manawadu, U. E., & Sugano, S. (2019). Human-centered intervention based on tactical-level input in unscheduled takeover scenarios for highly-automated vehicles. *International Journal of Intelligent Transportation Systems Research*, 1–10.

Kruglanski, A. W., Chernikova, M., Babush, M., Dugas, M., & Schumpe, B. (2015). The architecture of goal systems: Multifinality, equifinality, and counterfinality in means—End relations. *Advances in Motivation Science, 2*, 69–98.

Levi-Strauss, C. (1961). *A world on the wane* (J. Russell, Trans.). Criterion.

Marquis, C. (2003). The pressure of the past: Network imprinting in intercorporate communities. *Administrative Science Quarterly, 48*(4), 655–689.

Marx, K. (1867). *Das kapital* (B. Fowkes, Trans., 4 ed.). Capital.

Miller, D. D., & Brown, E. W. (2018). Artificial intelligence in medical practice: The question to the answer? *The American Journal of Medicine, 131*(2), 129–133.

Mischel, W., & Shoda, Y. (1998). Reconciling processing dynamics and personality dispositions. *Annual Review of Psychology, 49*(1), 229–258.

Mischel, W., & Shoda, Y. (2010). The situated person. In B. Mesquita, L. F. Barrett, & E. R. Smith (Eds.), *The mind in context* (pp. 149–173). Guilford Press.

Muth, J. F. (1961). Rational expectations and the theory of price movements. *Econometrica, 29*(3), 315–335.

Osoba, O. A., & Welser, W. (2017). *An intelligence in our image: The risks of bias and errors in artificial intelligence*. Rand Corporation.

Rawls, J. (1996). *Political liberalism*. Columbia University Press.

Rawls, J. (2001). *A theory of justice* (Revised ed.). Harvard University Press.

Riaz, F., Jabbar, S., Sajid, M., Ahmad, M., Naseer, K., & Ali, N. (2018). A collision avoidance scheme for autonomous vehicles inspired by human social norms. *Computers & Electrical Engineering, 69*, 690–704.

Scott, W. R. (2014). *Institutions and organizations: Ideas, interests, and identities* (4th ed.). Sage.

Scott, W. R., & Davis, G. F. (2007). *Organizations and organizing: Rational, natural and open system perspectives*. Pearson Education.

Sen, A. (1999). *Reason before identity*. Oxford University Press.

Shin, J., & Grant, A. M. (2020). When putting work off pays off: The curvilinear relationship between procrastination and creativity. *Academy of Management Journal* (online).

Smith, W. K., & Tushman, M. L. (2005). Managing strategic contradictions: A top management model for managing innovation streams. *Organization Science, 16*(5), 522–536.

Sull, D., & Eisenhardt, K. M. (2015). *Simple rules, how to thrive in a complex world*. Houghton Mifflin Harcourt.

Teece, D. J. (2014). The foundations of enterprise performance: Dynamic and ordinary capabilities in an (economic) theory of firms. *The Academy of Management Perspectives, 28*(4), 328–352.

Wood, W., & Rünger, D. (2016). Psychology of habit. *Annual Review of Psychology, 67*, 289–314.

Zhang, C., Vinyals, O., Munos, R., & Bengio, S. (2018). A study on overfitting in deep reinforcement learning. *arXiv preprint arXiv:1804.06893*.

Zuboff, S. (2019). *The age of surveillance capitalism: The fight for a human future at the new frontier of power*. Profile Books.

8

Learning

Intelligent agents perceive conditions and problems in the world, gather and process information, then generate solutions and action plans. And insofar as outcomes add to knowledge and associated procedures, agents learn. In this sense, much learning is responsive and adaptive, the result of cycles of problem-solving, action generation, performance evaluation, and updates. For human beings, inter-cyclical performance feedback is the primary source of such updates, unfolding as patterns of experience through time. Whereas for artificial agents, much learning is intensely intra-cyclical, meaning it occurs during action cycles, often in real time, mediated by rapid feedforward mechanisms. This is possible because artificial agents cycle far more rapidly and precisely, relative to most behavioral and mechanical processes. Indeed, artificial agents achieve unprecedented complexity and learning rates. Owing to this capability, artificial agents are increasingly important in practical problem-solving and process control, especially in environments where real-time adjustments are beneficial. The same capabilities will empower learning by digitally augmented agents.

Yet supervision is challenging here too. Human and artificial agents have noticeably different capabilities and potentialities in learning. On

© The Author(s) 2021
P. T. Bryant, *Augmented Humanity*, https://doi.org/10.1007/978-3-030-76445-6_8

the one hand, as stated above, artificial agents learn at high rates and levels of complexity and precision. They are hyperactive and hypersensitive in learning. For example, it only takes hours or days to train advanced artificial agents to high levels of expertise. On the other hand, human agents exhibit relatively sluggish learning rates and low levels of complexity. As any schoolteacher can attest, it takes years of incremental learning to educate a human being, and many never achieve expertise.

Clearly, these distinctions are like those in Chap. 7, regarding performance evaluation, which is no surprise because much learning is driven by performance feedback. Therefore, in both areas of functioning—the evaluation of performance and learning—human and artificial agents differ in terms of their processing rates, the complexity of processing, and primary mechanisms of updating. As noted, human learning is relatively sluggish, accrues in simpler increments, and mostly from inter-cyclical feedback, whereas, artificial agents are hyperactive and hypersensitive in learning, rapidly acquiring complex knowledge, including through feedforward mechanisms. It also follows that learning by augmented agents will exhibit the same potential distortions as the evaluation of performance. When combined in augmented agency, human and artificial learning can become divergent, dyssynchronous, and discontinuous, and thus ambiactive in terms of learning rates and levels of complexity.

Hence, learning by augmented agents can also skew, like the evaluation performance. Three patterns of distortion are possible. First, highly ambiactive learning will combine rapid, complex, artificial updates, with far slower, simpler, human updates. For example, digitalized learning systems cycle rapidly, shortening attention spans and compressing content, yet behavioral aspects of education and training require attentive dedication over long periods of time. In consequence, augmented learning could be overly divergent and hence dyssynchronous, discontinuous, and ambiactive. Second, in other situations, learning by augmented agents could be overly convergent and dominated by artificial processes which overwhelm or suppress human inputs. People would be increasingly reliant on digitalized procedures. Third, the opposite is also possible, in which learning is dominated by human myopia, bias, and idiosyncratic noise. Augmented learning would then reinforce and amplify erroneous priors. If any of these distortions occur, learning by augmented agents will be dysfunctional and often highly ambiguous and ambivalent.

8.1   Theories of Learning

Modern theories of learning emphasize the development of autonomous capabilities and reasoned problem-solving, rather than replication and rote memorization. Modern learning also highlights the role of experience and evaluative feedback. Via such mechanisms, agents develop capabilities, knowledge, and self-efficacy. For the same reasons, modern scholarship accords learning a major role in social development and human flourishing. Writing over a century ago, the founders of modern educational psychology espoused very similar principles (e.g., James, 1983; Pestalozzi, 1830). The enduring challenge is to explain and manage the deeper mechanisms of learning. For example, many continue to ask which aspects of learning are predetermined as natural priors, rather than resulting from experience and evaluative feedback. In other words, what is owing to nurture versus nature? And further, in which ways, and to what extent, can natural learning capabilities be enhanced, especially through structured experience and training? More recently, scholars also investigate how human and artificial agents best collaborate in learning (Holzinger et al., 2019).

Historical Debates

Once again, there is an impressive intellectual history. The ancient Greeks made important contributions which remain relevant today. For example, Plato argued that much knowledge was innate, bestowed by nature and inheritance. The challenge was then to release it. Whereas, Aristotle put more emphasis on nurture and learning through experience, and highlighted the role of memory in the absorption of such lessons (Bloch, 2007). Two thousand years later, Enlightenment scholars explored similar problems and solutions. John Locke (1979) explained learning in terms of the progressive encoding of new knowledge, initially onto a child's blank mind or *tabula rasa*. This complemented Rousseau's (1979) stronger emphasis on learning through the liberation of natural human curiosity, intuition, and inherent capability. Although, despite their differences, both Locke and Rousseau shared the modern view that all persons can learn and grow. Both elevated the status and potential of the autonomous, reasoning mind.

Later psychologists continued exploring the mechanisms of learning. In the mid-twentieth century, Skinner (1953) proposed a radical form of behaviorism, based on operant conditioning and reinforcement learning, driven by stimulus and response. However, then and now, critics view this approach as overly reductionist and materialistic. Most now agree that human beings are more intentional and agentic in learning. Not surprisingly, Bandura (1997) is among this community. He argues that humans learn much through experience and the modeling of behavior, which strengthen self-efficacy. Jerome Bruner (2004), another leading figure in educational and cognitive psychology, argues that human beings are embedded in culture and learn within it, as they compose and interpret narrative meaning. While for Gardner (1983), learning involves multiple intelligences, including rational and emotional, which engage a range of cognitive and affective functions. Different senses are engaged by these processes, including visual, auditory, and kinesthetic systems. To summarize, modern theories of learning emphasize the development of intelligent capabilities, the wider role of agentic functioning, the contextual nature of learning, and the importance of performance feedback.

Levels of Learning

At the individual level, contemporary theories of learning prioritize cognitive and affective factors, sensitivity to context, the value of experience, and the need for engagement, although, scholars have long disagreed about the mechanisms which explain these aspects of learning. For example, Chomsky (1957) argued for genetically encoded structures which scaffold grammar and the learning of language. Modern cognitive science was also emerging at the time, along with early computer science and neuroscience. Scientists started to explore the neurological systems which underpin human learning, akin to software architecture. Chomsky's work can be seen in this context. However, his critics saw innate knowledge structures as a throwback to scholastic conceptions, versus the liberation of autonomous mind (Tomalin, 2003). Many therefore resisted any notion of innateness, favoring fully developmental processes instead. For example, also around the mid-twentieth century, Piaget (1972) argued there are progressive stages of learning through childhood, corresponding to the development of the neurophysiological system, layering more complex concepts, relations, and logical structures. He argued that these structures were cumulative and contingent on early, albeit predictable developmental processes. Chomsky (Chomsky & Piatelli-Palmarini, 1980) disagreed and argued in response, that deep semantic structures emerge holistically, irrespective of context.

Recent research is more complex and nuanced, including Chomsky's (2014) own. No single psychological, behavioral, or neurophysiological model is fully explanatory. Like the agentic self generally, learning involves culture and context, physical embodiment, experience, and functional complexity. Evidence therefore points to more complex processes of development and procedures in learning (Osher et al., 2018). Reflecting this view, contemporary researchers focus on the variability of contexts and the way in which cognitive and neurophysiological processes interact in learning. They also investigate cognitive plasticity, with some arguing for relatively high levels of flexibility across the lifespan, while others are more conservative in this regard. Notably, recent studies show that the brain remains more plastic than previously thought (Magee & Grienberger, 2020). Many also highlight the importance of model-based learning, by which people master relatively complex patterns of thought and action in more holistic ways (Bandura, 2017). Formal education leverages model-based learning through experiential methods and problem-based instruction (Kolb & Kolb, 2009). The general trend is toward more complex models of learning which engage multiple components of the agentic system, including cognition, affect, different modalities, types of performance, feedback, and feedforward mechanisms (Bandura, 2007).

Similar ideas inform scholarship about learning at the group and collective levels. Theories emphasize functional complexity, integrating cognitive, affective, and behavioral factors, plus sensitivity to social, economic, and cultural contexts, and the potential to develop and grow over time (Argote et al., 2003). However, theories of collective learning also recognize strict limitations. In fact, they share many of the same concerns as individual level theories: cognitive boundedness, attentional deficits, poor absorptive capacity, persistent superstitious tendencies, plus myopias and biases (Denrell & March, 2001; Levinthal & March, 1993). Proposed solutions relate to the development of dynamic capabilities, flexible organizational design, the use of information technologies, and transactive memory systems, in which the storage and retrieval of knowledge are distributed among groups, allowing them to learn more efficiently (Wegner, 1995).

Procedural Learning

Another important strategy is procedural learning, which leverages individual habit and collective routine to reduce the processing load (Argote & Guo, 2016; Cohen et al., 1996). All agents benefit from acquiring less effortful learning procedures. The reader will recall that similar topics are discussed in Chap. 3, regarding agentic modality. In the earlier discussion, I review the debate about aggregation: whether collective routine emerges bottom-up, from the combination of individual habits, or functions top-down, from the devolution of social forms. In fact, the same questions arise for learning, namely, do collective learning routines emerge from the aggregation of individual learning habits, or vice versa? Some scholars privilege learning at the individual level and argue that routine learning is an aggregation of habit (Winter, 2013). In contrast, other scholars privilege the holistic origins of collective learning. From this alternative perspective, collective mind and action are the primitives of organizational learning, not the result of bottom-up aggregation. Therefore, questions of agentic modality come to the fore once again: does collective learning aggregate individual habits of learning, or vice versa? Similarly, do the limitations of collective learning reflect the aggregation of individual constraints, or do social and organizational factors impose limits on individual learning? Or perhaps all learning combines both types of constraint?

The solution to the aggregation question presented in Chap. 3, also applies to procedural learning. In this type of learning, many individual differences, such as personal encodings, beliefs, and goals, are downregulated and effectively latent, whereas shared characteristics, such as collective encodings, beliefs, and goals, are upregulated and active. Hence, learning habit and routine coevolve and are stored in individual and collective memory, integrated via common storage and retrieval processes. Collectives thus learn without activating significant differences, and without needing to aggregate such differences. Mischel and Shoda (1998) explain this is how cultural norms evolve, as common, recurrent psychological processes. Hence, routine learning is neither simply bottom-up nor top-down.

That said, collective agents also learn in nonroutine ways, from deliberate experimentation and risk-taking (March, 1991). Increasing environmental uncertainty and dynamism favor these approaches. In organizational life, this has led to an emphasis on continuous learning and methodologies which highlight feedforward processes, such as design thinking, lean startup methods, and agile software development (Contigiani & Levinthal, 2019). All provide ways for teams and organizations to learn in complex, dynamic environments, using intra-cyclical, adaptive means. Via such methods, agents learn despite high uncertainty, ambiguity, and accelerating rates of change. These methods contrast earlier, linear approaches toward learning, which emphasize slower, intercyclical feedback loops (e.g., Argyris, 2002).

Learning is therefore central to modern theories of agency. Apart from anything else, agents develop self-efficacy and capabilities through learning, by trying and sometimes failing, but succeeding often enough. In these respects, developmental learning exemplifies the vision of modernity, which is to nurture autonomous, intelligent agency, and thereby to increase the potential for human flourishing. Modern theories of learning prioritize the capability of agents to learn and grow, especially from the evaluation of performance. Digital augmentation promises major advances in all these functions.

8.2   Digitally Augmented Learning

Artificial agents are quintessentially problem-solving, learning systems. Some are fully unsupervised and self-generative, such as artificial neural networks and evolutionary machine learning. These agents are becoming genuinely creative, speculative, and empathic (Ventura, 2019). They also compose their own metamodels of learning, based on iterative, exploratory analysis, to identify the type of learning which fits best in any context. Artificial reinforcement learning is one recent development, which enables expert systems to learn rapidly from the ground up (Hao, 2019). Other agents are semi-supervised or even fully supervised. In these systems, models are encoded to guide learning and development. In all cases, artificial agents will employ a metamodel of learning, which is defined by its hyperparameters. Among other properties, hyperparameters will specify the potential categories and layers of learning, plus major mechanisms and cycle rates. These will include dynamic, intra-cyclical feedforward mechanisms, as well as longer inter-cyclical feedback processes.

Notably, artificial neural networks reflect the fundamental architecture of the human brain. This deep similarity between artificial and human agents facilitates close collaboration between them in augmented systems of learning (Schulz & Gershman, 2019). Both share similar features of neural architecture, and when joined, they can self-generate their own, augmented metamodels of learning. Some will be semi-supervised—such as Semi-Supervised Generative Adversarial Networks or SGANs—which might incorporate human values, beliefs, and commitments into metamodels of learning. Important applications already include artificial empathy and personality (Kozma et al., 2018). However, as in other areas of augmented functioning, the key challenge is to ensure effective supervision and metamodel fit.

Ambiactive Learning

Not surprisingly, the supervisory challenges of learning by augmented agents are like those of augmented evaluation of performance. To begin with, artificial agents will tend toward hypersensitive and hyperactive learning, meaning they quickly detect small degrees of variance, which trigger rapid, precise updates. As previously explained, this includes intra-cyclical feedforward mechanisms and entrogenous mediators, such as performative action generation. However, these mechanisms can lead to excessive updates and overlearning. Digitally augmented agents might learn too much and too often, thereby wasting time and resources. This could also produce ambiopic problem-solving, by encouraging sampling and searching too widely for problems and solutions, further and faster than required. For similar reasons, computer scientists research how to avoid unnecessary overlearning (Gendreau et al., 2013).

By comparison, human agents are frequently insensitive to outcome variance and sluggish in learning (Fiedler, 2012). Human learning is relatively simple and slow, compared to artificial agents. Significant problems therefore arise for augmented agents because they will combine sluggish, insensitive human updates, with hyperactive, hypersensitive artificial updates. As noted earlier, such learning could easily become dyssynchronous and discontinuous, and therefore ambiactive, meaning it simultaneously stimulates and dampens, different learning rates and levels of complexity and precision. For example, when people use online search about political matters, they trigger rapid, precise cycles of artificial processing, iterating rapidly to guide search and learning. But at the same time, the human agent may input inflexible, myopic priors as search terms. As a result, the learning process uses digitally augmented means to reinforce political bias. In fact, such learning is an expression of confirmation bias at digital scale and speed. And in most cases, it is highly ambiactive and dysfunctional. These scaling effects help to explain the rapid spread of fakery and falsehood on social networks.

As stated earlier, ambiactive learning also increases the risk of ambiguity and ambivalence, owing to its poorly synchronized, discontinuous nature. This is because ambiactive learning easily produces contradictory or incompatible beliefs, interpretations, and preferences. Granted, moderate degrees of ambiguity and ambivalence can be beneficial (Kelly et al., 2015; Rothman et al., 2017). They support creativity and enhance the robustness and flexibility of learning (March, 2010). But digitalization greatly amplifies these effects and extremes become more likely. Metamodel fit will be harder to achieve and sustain. Some people may become overly reliant on digitalized processes and incapable of autonomous, self-regulated learning.

Furthermore, excessive ambiguity and ambivalence can lead to cognitive dissonance and confusion, even triggering psychological and behavioral disorder, especially when they impact core beliefs and commitments (van Harreveld et al., 2015). Indeed, when ambiactive learning is extreme and widespread, people could lose a shared sense of reality, truth, and ethical norms (Dobbin et al., 2015; Hinojosa et al., 2016). Major consequences follow for digitally augmented communities and collectives. For without a shared sense of reality, truth, and right behavior, people are more vulnerable to deception and superstitious learning. Unable to discriminate real from fake, truth from falsehood, or right from wrong, they are more likely to be docile and rely on stereotypes.

Summary of Learning by Augmented Agents

Based on the foregoing discussion, we can now summarize the main features of learning by augmented agents. To begin with, it is important to recognize there are many potential benefits. Augmented agents acquire unprecedented capabilities to explore, analyze, generate, and exploit new knowledge and procedures. In many domains, significant benefits are already apparent. However, at the same time, the speed and scale of digitalization pose new risks. First, augmented agents risk discontinuous updating, because they might skew toward hypersensitive, complex artificial processing, while being relatively insensitive and simplified in human respects. Second, augmented agents risk dyssynchronous updating, because they might skew toward hyperactive, fast learning rates in artificial terms, while being sluggish in human terms. When combined, these divergent tendencies will produce ambiactive, dysfunctional learning, heightening the risks of ambiguity and ambivalence, and in extreme cases, superstitious learning and cognitive dissonance. The corresponding pattern of supervision is shown by segment 9 in Fig. 2.6. Alternatively, artificial agents might dominate learning and relegate human agency to the sidelines, like segment 7 in Fig. 2.6, or human agents could distort semi-supervised augmented learning, by importing myopia and bias, as in segment 3 of Fig. 2.6.

8.3 Illustrative Metamodels of Learning

This section develops illustrations of learning, showing premodern, modern, and digitalized metamodels. Like the earlier discussions of self-regulation in Chap. 6 and evaluation of performance in Chap. 7, the following illustrations highlight internal dynamics, especially the interaction of human and artificial agents in augmented systems. Also like the preceding two chapters, the following illustrations focus on contrasting processing rates and degrees of complexity, but this time in relation to the precision and rate of learning. Once again, the analysis highlights critical similarities and differences between human and artificial agents.

Lowly Ambiactive Modern Learning

Figure 8.1 illustrates the core features of a lowly ambiactive, modern system of learning. That is, learning in which agents are moderately assisted by technologies, and where updates are mainly from performance feedback, relatively synchronous and continuous. The horizontal axis depicts

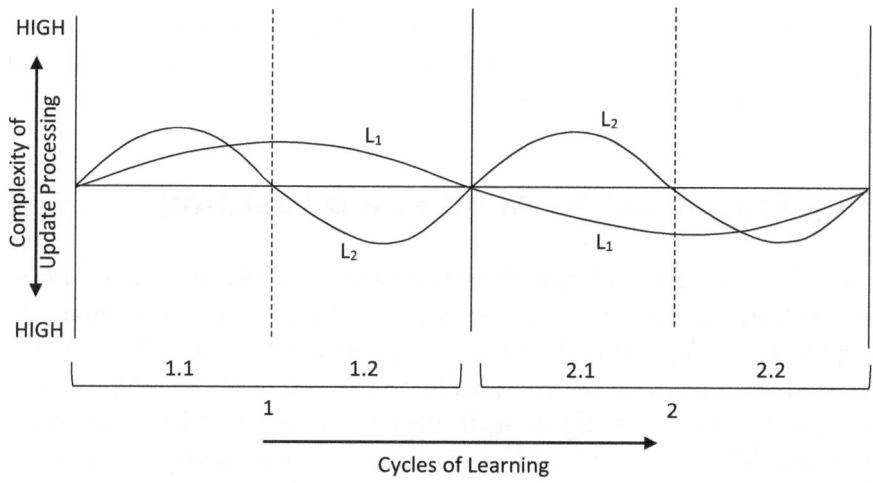

Fig. 8.1 Synchronous and continuous modern learning

two major cycles of learning labeled 1 and 2, which are further subdivided. These cycles encompass processes of feedback generation and subsequent updates to knowledge and procedures. The vertical axis illustrates the complexity and precision of learning updates, which range from low in the center to high in the upper and lower regions. The figure then depicts cycles for two metamodels of learning, shown by the curved lines labeled L_1 and L_2. Each depicts a full cycle of processing and updates. Metamodel L_2 cycles during each period at moderate complexity. However, metamodel L_1 cycles only once over periods 1 and 2, at a lower level of complexity. In this respect, L_1 represents a slower, simpler metamodel of learning, such as learning in premodern contexts. L_2 on the other hand, represents a faster, more complex metamodel of learning, which is typical of modernity. That said, every two cycles of L_2 are fully synchronized with one cycle of L_1, making the two metamodels moderately synchronous overall. In fact, L_2 is intra-cyclical relative to L_1, but assuming slower rates for both, synchronization is feasible. Both are of comparable complexity, and hence moderately continuous as well. As a combined system of learning, therefore, these two metamodels are moderately synchronous and continuous. Indeed, modern learning can be exactly like this. Traditional and cultural systems of learning, represented by L_1, are often moderately continuous and synchronized with technically assisted adaptive learning, represented by L_2. Overall, the scenario illustrated in Fig. 8.1 is therefore lowly ambiactive, often functional, and neither excessively ambiguous nor ambivalent.

Highly Ambiactive Augmented Learning

Next, Fig. 8.2 depicts highly dyssynchronous and discontinuous, ambiactive learning by augmented agents. Once more, the horizontal axis depicts two temporal cycles of learning, labeled 1.1 and 1.2, while the vertical axis illustrates the complexity and precision of learning updates, from low to high. The figure again depicts two metamodels of learning, this time labeled L_2 and L_3. The curved line L_2 is a modern metamodel of learning which assumes moderate technological assistance. L_3 depicts a fully digitalized, generative metamodel with a higher learning rate and greater complexity. Importantly, the two metamodels of learning are now poorly synchronized and connected. As the figure shows, they intersect in an irregular fashion. Updates are therefore dyssynchronous. Furthermore, the two metamodels exhibit different levels of complexity, which means updates will be discontinuous as well. If we now assume that both metamodels combine in one augmented agent—that is, the agent combines modern adaptive learning L_2 and digitalized generative learning L_3—overall learning will be dyssynchronous, discontinuous, ambiactive, probably dysfunctional, and highly ambiguous and ambivalent as well.

Consider the following example. Assume that L_2 in Fig. 8.2 represents a modern metamodel of adaptive learning, in which an aircraft pilot learns from practical performance feedback. Next, assume that L_3 represents the generative learning of an artificial avionic control system. Now assume that the pilot and avionic agent collaborate in flying an aircraft. Given their different modes of learning, they will update knowledge and procedures in a dyssynchronous and discontinuous fashion, reflecting the pattern in Fig. 8.2. Overall learning will be highly ambiactive, ambiguous, and ambivalent. Learning will likely be dysfunctional and, in this case, potentially disastrous. Indeed, aircraft have crashed for this reason (Clarke, 2019). Pilot training and artificial systems were poorly

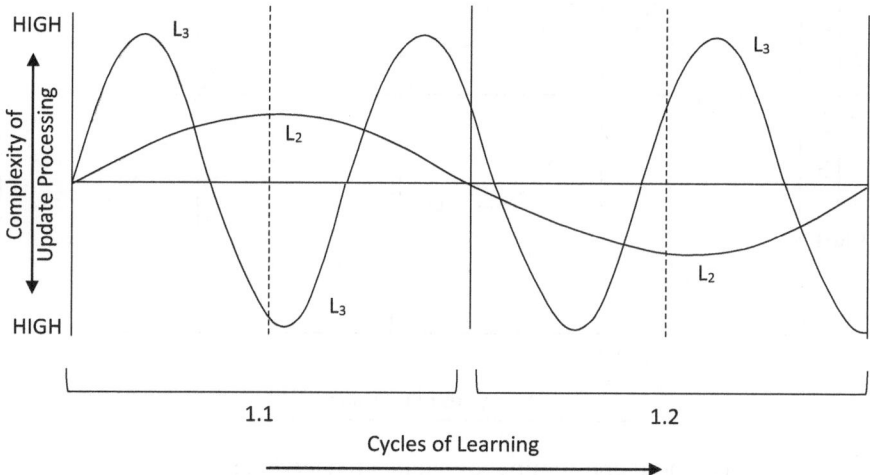

Fig. 8.2 Dyssynchronous and discontinuous augmented learning

coordinated and synchronized, and human and artificial agents failed to collaborate effectively. Pilots could not interpret or respond to the rapid, complex signals of the digitalized, flight control system. And the flight control system was insensitive to the needs and limitations of the pilots. The resulting accidents are tragic illustrations of ambiactive dysfunction. Similar risks are emerging in other expert domains, and many more instances are likely.

Lowly Ambiactive Augmented Learning

In contrast, Fig. 8.3 illustrates lowly ambiactive learning by an augmented agent. Digitalized learning is labeled L_4, and modern adaptive learning is again labeled L_2. As in the previous figures, the horizontal axis depicts temporal cycles, and the vertical axis again illustrates levels of complexity and precision. In contrast to the preceding figure, however, the two metamodels in Fig. 8.3 are now moderately synchronous and

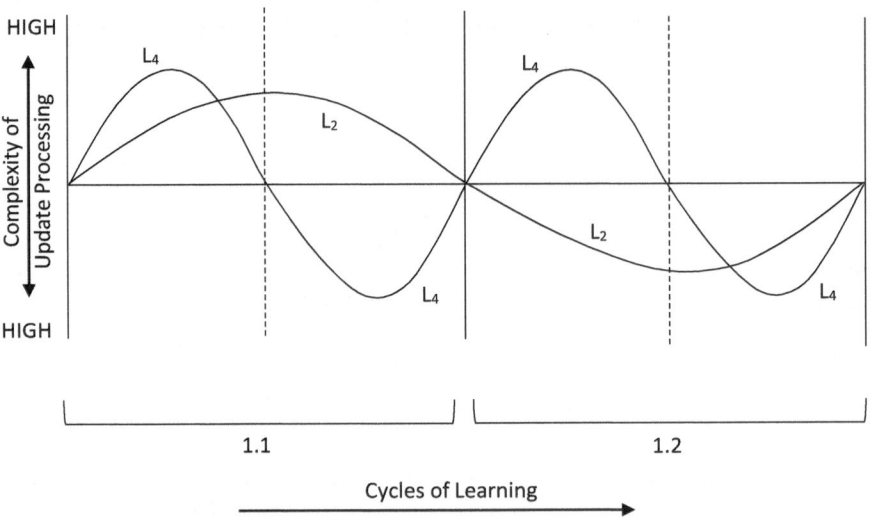

Fig. 8.3 Synchronous and continuous augmented learning

continuous. Cycles are better aligned, intersecting at the completion of major learning cycles, despite their different rates. Their levels of complexity are similar as well. By implication, collaborative supervision is strong. If we now assume that these two metamodels are combined in one augmented agent—that is, the agent combines modern adaptive learning L_2 and digitalized generative learning L_4—then overall learning will be lowly ambiactive, functional, and not significantly ambiguous or ambivalent. For the same reasons, entrogenous mediators will be adequately aligned, synchronous, and continuous as well. In summary, Fig. 8.3 illustrates a well-supervised system of learning by augmented agents, which is what engineers aspire to build (Chen et al., 2018; Pfeifer & Verschure, 2018). The supervision of learning achieves strong metamodel fit, in this case, with appropriately balanced learning rates and levels of complexity.

8.4   Wider Implications

Highly ambiactive learning therefore poses major risks for augmented agents: extreme ambiguity and ambivalence, incoherent and inconsistent updates, functional losses, and cognitive dissonance. Moving forward, researchers must develop methods of supervision which mitigate these risks and maximize metamodel fit. Fortunately, research has already begun. For example, dissonance engineering explicitly addresses these risks (Vanderhaegen & Carsten, 2017). It seeks to manage and supervise human-machine interaction in learning, and especially the risks of learning conflict within augmented systems. Other researchers are working to develop more empathic communication interfaces, to facilitate better human-machine communication in learning (Schaefer et al., 2017). However, we have yet to see comparable research efforts in the social and behavioral sciences. Following sections highlight some of the major challenges and opportunities, in this regard.

Divergent Capabilities

Many of these problems arise, because the speed and scale of digital innovation are outpacing human absorptive capacities and traditional methods of learning. Most human learning is gradual, cycles relatively slowly, responding to inter-cyclical performance feedback. Knowledge is absorbed incrementally, often vicariously. Theories therefore assume broadly adaptive processes, driven by experience and performance feedback, iterating in punctuated gradualism over time. Moreover, owing to limited capabilities and behavioral contingency, human learning is often incomplete. In contrast, artificial learning is increasingly powerful, fast, and self-generative. Indeed, for today's most advanced artificial agents, it may only take minutes or hours to perform complex learning tasks, which no human agent could ever complete. Compounding the challenge, artificial feedforward mechanisms are largely inaccessible to consciousness, given the relatively sluggish, insensitive nature of human monitoring. When combined, these divergent capabilities produce novel risks for augmented agents. Artificial processes could race ahead, while humans constantly struggle to keep up. Human attention spans may continue to shrink, while digitalized content is compressed and commoditized (Govindarajan & Srivastava, 2020). Recall that similar effects drive entrogenous divergence in Fig. 2.4. Learning could be simultaneously adaptive and generative, fast and slow, over-complete and incomplete. At the same time, human supervision could import erroneous myopia, bias, and noise, and digital augmentation will reinforce and amplify these limitations. In these situations, learning will be dysfunctional and lead to functional losses.

Existing theories of learning are ill-equipped to conceptualize and explain these risks. Theories typically assume the gradual, progressive absorption of knowledge and skills (Lewin et al., 2011). Myopic learning, insensitivity to feedback, and sluggishness are the main foci of scholarly attention, which makes perfect sense in a pre-digital world. For the same reasons, the risks of overlearning and over-absorption receive little attention. It is no surprise, therefore, that existing theories are ill-equipped to explain the effects of digitally augmented hyperopia, hyperactivity, and hypersensitivity, and the resulting risks of dyssynchronous, discontinuous updating, and ambiactive learning. In fact, to date, most of these risks are not even conceptualized in theories of learning.

The opposite is true for research in computer science and artificial intelligence. In these domains, overlearning, hypersensitivity, and hyperactivity already receive significant attention (Panchal et al., 2011), which is fully explicable because the risks are clear and growing. They interrupt and confuse artificial learning, leading to functional losses. Consider the control of autonomous vehicles once again. In an emergency, the artificial components of the augmented system should prioritize rapid approximate learning over slow exact learning. For example, there is no need to distinguish a pedestrian's age or gender before stopping to avoid a collision (Riaz et al., 2018). Overlearning would delay action and be disastrous. Comparable trade-offs will arise in other augmented domains, including the piloting of aircraft, as discussed earlier. Supervised trade-offs will be critical, balancing the complexity of updates versus learning rates, given the context and goals. If well supervised, augmented agents can enjoy significant benefits. They will incorporate the best of human experience and commitment, and the best of artificial computation and discovery.

Furthermore, digitally augmented agents will recompose metamodels of learning in real time, to maintain metamodel fit as contexts change. To do so, they will rely heavily on entrogenous mediators, namely intelligent sensory perception, performative action generation, and contextual learning. Every phase of learning will be intelligent and generative, rather than procedural and incremental. In this fashion, digitally augmented learning enables adaptive learning by design. But this raises additional questions. How much of augmented learning will be accessible to human consciousness and supervision, or will it be an opaque product of artificial intelligence? And if the latter proves true, could this lead to a new type of superstitious learning, in which humans absorb outcomes without understanding how or why they came about? In fact, we already see some evidence of this, in the opacity of deep learning and artificial neural networks. These systems are widely applied in augmented systems, but important features can remain hidden and not explainable, even to the developers (Pan et al., 2019).

Problematics of Learning

When viewed collectively, these developments signal a major change in the problematics of learning. To begin with, recall that modernity problematizes the following: how and to what degree, can human beings transcend their natural limits through learning, to be more fully rational, empathic, and fulfilled? Digitalization prompts additional questions. To begin with, how can human beings collaborate closely with artificial agents in learning while remaining genuinely autonomous in reasoning, belief, and choice? Relatedly, how can humans absorb digitally augmented learning while preserving their natural intuitions, instincts, and commitments? And how can digitally augmented institutions and organizations, conceived as collective agents, fully exploit artificial learning while avoiding extremes of digitalized determinism? Also, how can humanity ensure fair access to the benefits of digitally augmented learning and not allow it to perpetuate inequality, discrimination, and injustice? Finally, how will human and artificial agents trust and respect each other in collaborative learning, despite their different levels of capability and potentiality? The metamodels and mechanisms described here can help guide research into these questions.

References

Argote, L., & Guo, J. M. (2016). Routines and transactive memory systems: Creating, coordinating, retaining, and transferring knowledge in organizations. *Research in Organizational Behavior, 36*, 65–84.

Argote, L., McEvily, B., & Reagans, R. (2003). Managing knowledge in organizations: An integrative framework and review of emerging themes. *Management Science, 49*(4), 571–582.

Argyris, C. (2002). Double-loop learning, teaching, and research. *Academy of Management Learning & Education, 1*(2), 206–218.

Bandura, A. (1997). *Self-efficacy: The exercise of control.* W.H.Freeman and Company.

Bandura, A. (2007). Reflections on an agentic theory of human behavior. *Tidsskrift-Norsk Psykologforening, 44*(8), 995.

Bandura, A. (Ed.). (2017). *Psychological modeling: Conflicting theories.* Transaction Publishers.

Bloch, D. (2007). *Aristotle on memory and recollection: Text, translation, interpretation.* Brill.

Bruner, J. (2004). Life as narrative. *Social Research, 71*(3), 691–710.

Chen, J. Y. C., Lakhmani, S. G., Stowers, K., Selkowitz, A. R., Wright, J. L., & Barnes, M. (2018). Situation awareness-based agent transparency and human-autonomy teaming effectiveness. *Theoretical Issues in Ergonomics Science, 19*(3), 259–282.

Chomsky, N. (1957). *Syntactic structures.* Mouton & Co.

Chomsky, N. (2014). *The minimalist program.* MIT Press.

Chomsky, N., & Piattelli-Palmarini, M. (1980). *Language and learning: The debate between Jean Piaget and Noam Chomsky.* Harvard University Press.

Clarke, R. (2019). Why the world wants controls over artificial intelligence. *Computer Law & Security Review, 35*(4), 423–433.

Cohen, M. D., Burkhart, R., Dosi, G., Egidi, M., Marengo, L., Warglien, M., & Winter, S. (1996). Routines and other recurring action patterns of organizations: Contemporary research issues. *Industrial and Corporate Change, 5*(3), 653–688.

Contigiani, A., & Levinthal, D. A. (2019). Situating the construct of lean start-up: Adjacent conversations and possible future directions. *Industrial and Corporate Change, 28*(3), 551–564.

Denrell, J., & March, J. G. (2001). Adaptation as information restriction: The hot stove effect. *Organization Science, 12*(5), 523–538.

Dobbin, F., Schrage, D., & Kalev, A. (2015). Rage against the iron cage: The varied effects of bureaucratic personnel reforms on diversity. *American Sociological Review, 80*(5), 1014–1044.

Fiedler, K. (2012). Meta-cognitive myopia and the dilemmas of inductive-statistical inference. *Psychology of Learning & Motivation, 57*, 1–55.

Gardner, H. (1983). *The theory of multiple intelligences.* Heinemann.

Gendreau, M., Hyde, M., Kendall, G., Ochoa, G., Ozcan, E., & Qu, R. (2013). Hyper-heuristics: A survey of the state of the art. *Journal of the Operational Research Society, 64*(12), 1695–1724.

Govindarajan, V., & Srivastava, A. (2020). What the shift to virtual learning could mean for the future of higher ed. *Harvard Business Review, 31*, 1–4.

Hao, K. (2019, January 25). We analyzed 16,625 papers to figure out where ai is headed next. *MIT Technology Review* (online).

Hinojosa, A. S., Gardner, W. L., Walker, H. J., Cogliser, C., & Gullifor, D. (2016). A review of cognitive dissonance theory in management research: Opportunities for further development. *Journal of Management, 43*(1), 170–199.

Holzinger, A., Plass, M., Kickmeier-Rust, M., Holzinger, K., Criayan, G. C., Pintea, C.-M., & Palade, V. (2019). Interactive machine learning: Experimental evidence for the human in the algorithmic loop. *Applied Intelligence, 49*(7), 2401–2414.

James, W. (1983). *Talks to teachers on psychology and to students on some of life's ideals* (Vol. 12). Harvard University Press.

Kelly, R. E., Mansell, W., & Wood, A. M. (2015). Goal conflict and well-being: A review and hierarchical model of goal conflict, ambivalence, self-discrepancy and self-concordance. *Personality and Individual Differences, 85*, 212–229.

Kolb, A. Y., & Kolb, D. A. (2009). The learning way: Meta-cognitive aspects of experiential learning. *Simulation & Gaming, 40*(3), 297–327.

Kozma, R., Alippi, C., Choe, Y., & Morabito, F. C. (Eds.). (2018). *Artificial intelligence in the age of neural networks and brain computing*. Academic Press.

Levinthal, D. A., & March, J. G. (1993). The myopia of learning. *Strategic Management Journal, 14*(S2), 95–112.

Lewin, A. Y., Massini, S., & Peeters, C. (2011). Microfoundations of internal and external absorptive capacity routines. *Organization Science, 22*(1), 81–98.

Locke, J. (1979). *The Clarendon edition of the works of John Locke: An essay concerning human understanding*. Clarendon Press.

Magee, J. C., & Grienberger, C. (2020). Synaptic plasticity forms and functions. *Annual Review of Neuroscience, 43*, 95–117.

March, J. G. (1991). Exploration and exploitation in organizational learning. *Organization Science, 2*(1), 71–87.

March, J. G. (2010). *The ambiguities of experience*. Cornell University Press.

Mischel, W., & Shoda, Y. (1998). Reconciling processing dynamics and personality dispositions. *Annual Review of Psychology, 49*(1), 229–258.

Osher, D., Cantor, P., Berg, J., Steyer, L., & Rose, T. (2018). Drivers of human development: How relationships and context shape learning and development. *Applied Developmental Science, 24*(1), 6–36.

Pan, Z., Yu, W., Yi, X., Khan, A., Yuan, F., & Zheng, Y. (2019). Recent progress on generative adversarial networks (GANs): A survey. *IEEE Access, 7*, 36322–36333.

Panchal, G., Ganatra, A., Shah, P., & Panchal, D. (2011). Determination of over-learning and over-fitting problem in back propagation neural network. *International Journal on Soft Computing, 2*(2), 40–51.

Pestalozzi, J. H. (1830). *Letters of Pestalozzi on the education of infancy: Addressed to mothers*. Carter and Hendee.

Pfeifer, R., & Verschure, P. (2018). The challenge of autonomous agents: Pitfalls and how to avoid them. In *The artificial life route to artificial intelligence* (pp. 237–264). Routledge.

Piaget, J. (1972). Development and learning. *Readings on the Development of Children*, 25–33.

Riaz, F., Jabbar, S., Sajid, M., Ahmad, M., Naseer, K., & Ali, N. (2018). A collision avoidance scheme for autonomous vehicles inspired by human social norms. *Computers & Electrical Engineering, 69*, 690–704.

Rothman, N. B., Pratt, M. G., Rees, L., & Vogus, T. J. (2017). Understanding the dual nature of ambivalence: Why and when ambivalence leads to good and bad outcomes. *Academy of Management Annals, 11*(1), 33–72.

Rousseau, J.-J. (1979). *Emile or on education* (A. Bloom, Trans.). Basic Books.

Schaefer, K. E., Straub, E. R., Chen, J. Y. C., Putney, J., & Evans, A. W. (2017). Communicating intent to develop shared situation awareness and engender trust in human-agent teams. *Cognitive Systems Research, 46*, 26–39.

Schulz, E., & Gershman, S. J. (2019). The algorithmic architecture of exploration in the human brain. *Current Opinion in Neurobiology, 55*, 7–14.

Skinner, B. F. (1953). *Science and human behavior*. The Free Press.

Tomalin, M. (2003). Goodman, Quine, and Chomsky: From a grammatical point of view. *Lingua, 113*(12), 1223–1253.

van Harreveld, F., Nohlen, H. U., & Schneider, I. K. (2015). The abc of ambivalence: Affective, behavioral, and cognitive consequences of attitudinal conflict. In *Advances in experimental social psychology* (Vol. 52, pp. 285–324). Elsevier.

Vanderhaegen, F., & Carsten, O. (2017). Can dissonance engineering improve risk analysis of human-machine systems? *Cognition, Technology & Work, 19*(1), 1–12.

Ventura, D. (2019). Autonomous intentionality in computationally creative systems. In *Computational creativity* (pp. 49–69). Springer.

Wegner, D. M. (1995). A computer network model of human transactive memory. *Social Cognition, 13*(3), 319–339.

Winter, S. G. (2013). Habit, deliberation, and action: Strengthening the microfoundations of routines and capabilities. *The Academy of Management Perspectives, 27*(2), 120–137.

9

Self-Generation

Modern persons are most fulfilled when they freely choose who to be and become in the world. In other words, they flourish best through autonomous self-generation, which is to manage one's own development and reproduction without external direction. At the individual level, people self-generate career paths, social identities, and autobiographical narratives (McAdams, 2013). Options for doing so are found in culture and community, which provide choice sets of possible selves and life courses. Similarly, groups and larger collectives self-generate through organized goal pursuit and the composition of shared narratives (Bruner, 2002). Options at this level emerge from culture, social ecology, and history. These choice sets also comprise metamodels of self-generative potentiality, that is, related sets of self-generative models. Any choice will therefore instantiate one or other agentic metamodel.

Self-generative potentiality also varies from culture to culture, between social-economic groups, and across historical periods. Regarding the past, as previous chapters explain, in premodern contexts, agentic potentiality was tightly constrained. Metamodels of agency were relatively fixed and stable and provided few degrees of self-generative freedom. For most people in premodernity, life was dominated by tradition and templates for

© The Author(s) 2021
P. T. Bryant, *Augmented Humanity*, https://doi.org/10.1007/978-3-030-76445-6_9

survival. Living a good life meant having physical security, food and shelter, family continuity, and the replication of communal rituals and norms. Similar principles are applied at the collective level. Social organization was stable and patriarchal. Collective self-generation referenced embedded norms and established orders. Indeed, these are the core features of the replicative, agentic metamodel which dominated during premodernity.

By contrast, during the modern era, self-generative potentiality expanded greatly, at least for many. As capabilities and endowments increased, people enjoyed greater degrees of freedom and choice, to develop as autonomous, self-efficacious agents. In many societies, cultural norms have shifted in the same direction, to encourage personal ambition and mobility. Reflecting such freedom, the modern period is characterized by self-generative possibility. It aspires to liberate human potential, transcending the premodern focus on survival and fate. Modernity tells a story of progress, reasoning mind, scientific discovery, and innovation, all dedicated to the "social conquest of earth" (Wilson, 2012). And to be sure, progressive social policies and economic growth have expanded self-generative capability and potentiality. Technological innovation, improvements in education, public health, participatory government, and free market economies have combined to lift many (though not all) from historic deprivation and ignorance. For example, the career path of entrepreneurship is now a well-established option in contemporary societies (McAdams, 2006). It incorporates values of autonomy and exploration, a preference for risk-taking, creativity, and organized goal pursuit—all qualities which exemplify the adaptive, agentic metamodel of modernity. Production and consumption have also grown, leading to a predictable emphasis on the acquisition of goods and services, and the enjoyment of their utility.

Nevertheless, self-generation often falls short of aspirations, owing to enduring constraints and deficits. To begin with, options remain limited for many. Survival may be the best a person or community can hope for. What is more, self-generation can also disappoint in relatively abundant environments. Even if better, more varied self-generative options emerge, they might prove difficult to realize, because agents are incapable of choice and lack conversion capabilities, being the capabilities required to exploit the opportunities one has (Sen, 2000). Hence, owing to increasing complexity and limited capabilities, people neither discriminate between options nor convert them into reality. The expansion of potentiality overwhelms them. In these situations, many rely on social docility instead. They adopt the career and life path recommended by their community or family. Although that said, this kind of docility is often satisfying, especially in relatively munificent societies. Living a standard life in a plentiful world can be fulfilling enough.

Contemporary digitalization amplifies these opportunities and challenges. For example, at the individual level, digitalization provides new ways for people to curate and share memories, form new relationships, and choose alternative identities and futures. Digitalization also creates fresh opportunities at the collective level, to organize, collaborate, pursue common goals, and compose new narratives. Artificial agency points in the same direction. Particularly, today's most advanced systems are fully self-generative and globally connected. In these respects, human and artificial agents are increasingly compatible, as intelligent self-generative agents. A pluralistic world of augmented potentiality is fast emerging. However, at the same time, digitalization amplifies the dilemmas of munificence described earlier. Presented with a rapidly expanding range of self-generative options, many are unprepared, resistant, or overwhelmed, by the range and complexity of choice. They resist, delay, or retreat from digitalized, self-generative options (see Kozyreva et al., 2020).

Ironically, therefore, and in contrast to earlier periods, digitally augmented self-generation may disappoint because of too many opportunities and resources, rather than too few. To be sure, self-generative potentiality will increase, but if human capabilities lag, the freedom to choose will decline. In fact, recent studies report such effects (e.g., Scott et al., 2017). They show that digitally augmented self-generation can skew toward such extremes. People might resist digitalization on some dimensions, feel blocked and incapable in other ways, or retreat to their priors, while others surrender to digital determination (Collins, 2018). For example, in curating an online persona, some people deliberately avoid information about alternative life choices, yet struggle to search the sources they trust, and therefore rely on artificial agents to determine their choices. In this fashion, digital augmentation might narrow and distort self-generation.

9.1   Self-Generative Dilemmas

In fact, self-generative trade-offs are the norm. All agents compromise to some degree, as they balance self-generative freedom with the need for coherence and control (Bandura, 1997; Schwartz, 2000). One common strategy is to limit the range of options under consideration. As I explained earlier, people often simplify choice by relying on docility within the social world (Ryan & Deci, 2006). They defer to culture and convention, rather than autonomous reflection, when making self-generative choices. They adopt myopic life paths and focus on singular domains of being and doing. To be sure, myopia and social docility simplify choice (Bargh & Williams, 2006). Myopic choices are typically clear and predictable. And docility allows people to find psychosocial meaning and continuity within culture (McAdams, 2001). They choose from a preexisting set of possible futures, confident in their meaning and feasibility. By choosing mimetic life paths, therefore, agents can self-generate with a modest sense of autonomy, while securing coherence and consistency.

Sources of Disturbance

However, self-generative coherence is easily disturbed, especially if the choice set suddenly contracts or expands. Regarding the contraction of choice, a sudden loss of resources or social order will reduce the range of self-generative options. Disease, social disorder, or economic depression may strike, and sometimes all three, as in times of global pandemic. When such events occur, there are fewer opportunities and degrees of freedom for self-generation. Potentiality shrinks and lives are disrupted. In contrast, regarding the sudden expansion of choice, a rapid increase in resources, capabilities, or endowments will enhance options for self-generation. For example, a person may unexpectedly inherit a fortune, or be transported to an abundant environment, or gain access to extraordinary knowledge and capabilities. Similarly, a community might discover vast, untapped resources. Self-generative choice sets rapidly expand.

However, plentiful choice is unusual and presents different challenges. As stated above, some agents struggle to appreciate an expanded range of possibilities and find it hard to discriminate and order preferences. And even if they can choose, they may fail to realize their choice, owing to inadequate conversion capabilities and lack of requisite resources, especially when self-generative options are novel and complex. Hence, people are myopic and simplify choice. They make singular, predictable life choices. Opportunities are missed, and sometimes intentionally avoided (Bandura, 2006). In any case, self-generative abundance is exceptional. For most individuals and communities, the opposite is true. They endure deprivation as a permanent condition and have few self-generative options at the best of times (Sen, 2000). Not surprisingly, therefore, the dilemmas of munificence are rarely studied, apart from some fictional accounts (e.g., Forster, 1928; Huxley, 1998), and almost never treated as problematic. Rather, scholarly attention rightly focuses on persistent limitations and deprivation.

Digital Augmentation of Self-Generation

Digitalization promises a qualitative shift in this regard. Quite simply, it affords more options for being, doing, and becoming. By leveraging digitalized capabilities, augmented agents will be able to combine different modes of action and becoming, self-generating dynamically in real time. Consider clinical medicine once again. In this domain, augmented human-machine agents (clinicians and computers) will combine empathy and personality, associative and speculative analysis, clinical expertise and robotic capability, plus predictive scenario modeling, all simultaneously in real time. Working together, they will take patient care to a new level. In this fashion, digitalization will enable more dynamic, flexible self-generation by clinicians, as people and professionals. More generally, it will allow augmented agents to function effectively across multiple modes of being and doing (see Chen & Dalmau, 2005), that is, to collaborate in ambidextrous self-generation.

Digitalization therefore continues the narrative of modernity, toward richer self-generative capability and potentiality, but now at great scale and speed (Bandura, 2015). In fact, the digital augmentation of self-generation will constitute a historic transformation, at least for many, toward self-generative abundance and ambidexterity, in contrast to historic patterns of limited, singular modes of activity and self-generation. Already, digital networks allow people to adopt new modes of action and compose alternative narratives within virtual worlds. Similarly, online communities proliferate, while digital platforms support innovative social and organizational forms (Baldwin, 2012). Moreover, future innovations will accelerate these trends. Even at the bottom of the socioeconomic pyramid, digitalization allows a growing number of people to aspire to forms of life and action which were previously inconceivable (Mbuyisa & Leonard, 2017).

Nevertheless, as noted earlier, some people will retreat, resist, feel blocked, or simply be incapable of embracing new possibilities. Indeed, most people are poor at combining new and different modes of action. For example, many cannot synthesize associative and calculative intelligence, nor can they combine creativity and computation (see Malik et al., 2017). Similarly, they struggle to absorb alternative modes of being and doing in social life. Most people are not ambidextrous in these respects. In fact, this limitation is reflected in the classic metamodel of industrial modernity: the strict division of labor, singular domains of training and efficacy, and path-dependent careers. For this reason, contemporary educational and training programs try to develop ambidextrous capabilities, especially in managing opportunity and innovation (O'Reilly & Tushman, 2013).

Other people will retreat or resist the digital augmentation of self-generation, especially those who are deeply committed to cultural traditions or have inflexible assumptions about the ideal self. For these people, digital augmentation will not expand self-generative potentiality. It will reinforce myopic priors instead. For example, studies document the proliferation of online xenophobia against alternative life choices (Chetty & Alathur, 2018). At the opposite extreme, some people could overly relax and abandon their prior commitments. Instead of maintaining cultural norms and values, and seeking to own their own choices, they may surrender to artificial control and become digitally docile. Their domains of action, even careers and life paths, will be determined by artificial sources. Risks therefore emerge at every extreme. Neither retreat, resistance, blockage, nor surrender are effective responses to the digital augmentation of self-generation.

Dilemmas of Augmentation

New dilemmas thus emerge for augmented agents, as they seek to self-generate. On the one hand, artificial agents are increasingly self-generative, able to combine different modes of action and intelligence in real time, far beyond the reach of human capabilities and consciousness. While on the other hand, humans are typically sluggish and myopic in self-generation and tend toward singular modes of being and doing. Therefore, when human and artificial agents collaborate as augmented agents, they bring different self-generative strengths and weaknesses. If poorly supervised, the combined system could be singular and path dependent in human respects, but variable and dynamic in artificial terms. This will result in divergent, distorted patterns of self-generative ambidexterity. Agents will combine singular, exploitative modes of human self-generation, with flexible, exploratory modes of artificial self-generation. Alternatively, one agent might dominate the other and self-generation is highly convergent, for example, when people surrender to artificial determination.

This presents another supervisory challenge for augmented agents. They must find an appropriate ambidextrous balance, that is, combining human and artificial modes of self-generation to maximize metamodel fit. If supervision is poor, however, the result will be dysfunctional divergence or convergence. Consider the following example. Assume that some years ago, a woman or man trained to be a schoolteacher and learned traditional pedagogical methods. A predictable life course lay ahead. However, more recently, rapid digitalization, pandemic risks, and other social developments, require the teacher to master digitally augmented techniques and tools. In other words, the teacher must now collaborate in ambidextrous self-generation. However, she or he may resist or feel blocked, and default to prior knowledge and procedures. The risk, therefore, is that human and artificial self-generation will be divergent and dysfunctional. Digitally augmented self-generation would be a distorted form of ambidexterity, in which both agents are likely to obstruct each other. In fact, recent studies show that this is already happening (e.g., Salmela-Aro et al., 2019).

These dilemmas suggest a major shift in the problematics of self-generation. As noted above, modern scholarship rightly focuses on the persistent deprivations and limitations which constrain self-generation (Sen, 2017a). However, in a highly digitalized world, the problematics of self-generation expand. In addition to overcoming limits and deprivation, humanity must learn to appreciate and absorb digitally augmented potentiality. Lifelong learning and self-regeneration will become the norm. New questions therefore arise: how can human and artificial agents collaborate in dynamic self-generation, learning to be jointly ambidextrous in this regard, while ensuring human coherence and continuity; and what will count as well-being and flourishing, in a digitally augmented world?

Summary of Augmented Self-Generation

In summary, whether for good or ill, digitalization is transforming established patterns of self-generation. On the one hand, artificial self-generation is increasingly exploratory and autonomous, as artificial agents compose and recompose themselves. By incorporating these capabilities, augmented agents will be capable of multiple modes of being, doing, and becoming. They will be ambidextrous in this regard, combining both human and artificial modes of self-generation. On the other hand, however, human agents naturally possess limited capabilities and often remain committed to exploiting singular narratives and traditional life paths. They are persistently non-ambidextrous, unless trained to be otherwise. When both types of agent combine in augmented agency, the result could be self-generative divergence or convergence. Regarding divergence, human self-generative functioning will combine and conflict with artificial functioning (e.g., Levy, 2018). And regarding convergence, some people will either overtake or surrender to artificial self-generation. The digital augmentation of self-generation therefore focuses this book's core question: how to be and remain agentic in a digitalized world? The challenge is to supervise self-generation in ways which exploit new capabilities and potentialities, while respecting human choices and commitments.

9.2   Illustrations of Self-Generation

The preceding argument identifies the following principles. Human and artificial agents are situated, complex, open, and adaptive systems. Both exhibit varying degrees of self-generative capability and potentiality. However, humans have limited capability to discriminate, choose, and explore new self-generative options. Instead, they often exploit singular and predictable modes of self-generation, while artificial agents are increasingly dynamic and exploratory. In consequence, many humans will retreat, resist, feel blocked, or simply surrender, in response to the digital augmentation of self-generation. In these situations, digitalization will produce distorted forms of self-generative ambidexterity. Assuming these principles, the following sections illustrate major scenarios of self-generation, including the new patterns emerging in today's augmented world. The first illustration shows the baseline of modernity.

Self-Generation in Modernity

As earlier sections explain, modernity aspires to develop autonomous reasoning persons who can self-generate their own life path. Contemporary educational and behavioral interventions exemplify these aspirations, as do modern institutions and organizations (Scott, 2004). Figure 9.1 illustrates the self-generative metamodels within such a world. The figure focuses on the core challenge discussed in the previous section, namely, the capability of agents to discriminate and choose between self-generative options—the major risk being that people discriminate poorly, often resist or surrender, and fail to maximize choice. To capture these effects, the figure compares the complexity of self-generative metamodels to the degree of discriminate ranking between them. The figure further assumes capabilities at level L_2, with a moderate level of technological assistance. It also assumes that the more complex the self-generative choice set, or metamodel of self-generation, the less discriminated it is likely to be, and vice versa.

The figure then depicts four metamodels of self-generative choice. Quadrant 1 combines complex self-generative models, with highly discriminate ranking of them. This implies that agents can make a best

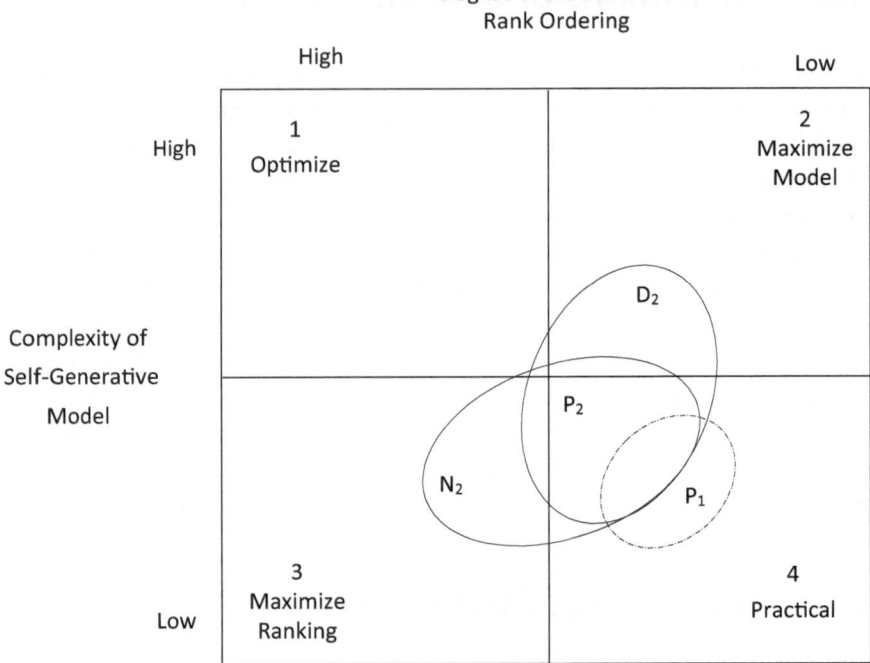

Fig. 9.1 Modern self-generation

choice about a complex model. Hence, these choices are optimizing. But they demand strong ambidextrous, self-generative capabilities, which enable agents to discriminate and combine multiple, complex options. Second, quadrant 2 combines complex self-generative models, with less discriminate ranking. Such choices will be maximizing. Agents will incompletely rank complex options and choose one which is no worse than alternatives. This scenario assumes moderate ambidextrous capabilities and is more feasible in this respect. Indeed, it accords with observed reality: people often choose no worse versions of complex life paths—for example, adopting an entrepreneurial career, in which options are complex and hard to rank. Next, quadrant 3 combines less complex self-generative models with highly discriminate ranking of them. Such choices will also be maximizing, owing to the almost complete rank ordering of less complex options. This metamodel also assumes moderate ambidextrous capabilities. And once again, it accords with observation: many people choose the best version of a simpler life path, for example, striving to achieve elite career status in a highly regulated community or profession. Finally, quadrant 4 combines less complex self-generative models with less discriminate ranking. These choices will be practical, meaning they are feasible and likely to succeed, and adequate for being a self-generative agent in the world. Not surprisingly, this metamodel assumes lesser capabilities and is therefore very feasible. Arguably, many individuals and collectives exhibit this type of self-generation: choosing a no worse version of a simpler life path. Making routine, mimetic choices in a modern world and being adequately fulfilled by doing so.

Figure 9.1 also shows further details. Different metamodels of self-generation, or model choice sets, are shown by the oval shapes N_2, D_2, and P_2. First, it is important to note, that these metamodels do not encompass much of the optimizing quadrant 1. Such choices are ideal and inspirational, but difficult to rank and realize, owing to their extreme complexity and the required level of discrimination. Second, N_2 is primarily overlapping quadrant 3, which combines simplified models, with highly discriminate ranking of them. These options will be maximizing, with respect to the complete ordering of simplified, self-generative models. This metamodel therefore assumes moderate capabilities, at best. It is also more normative and calculative, for example, by planning to achieve elite status within a regulated community or profession. Hence, the symbol N is employed. Third, the metamodel D_2 is primarily overlapping quadrant 2, which combines complex self-generative models, with partial, less discriminate ranking. It also assumes moderate capabilities. These options are maximizing, with respect to the incomplete rank ordering of complex models. Hence, the symbol D is used, and the self-generative options in D_2 are more descriptive, intuitive, associative, and harder to discriminate—for example, choosing an entrepreneurial career and life path. And fourth, the metamodel P_2 largely overlaps quadrant 4, which illustrates a practical self-generated life in the modern world, following a narrow, routine path with modest expectations or aspirations, which is adequate, feasible, and hence the most frequent choice.

Note that the figure also shows another scenario labeled P_1. This indicates the practical self-generative choices of a premodern world. Clearly, P_1 is even less complex and discriminated than P_2, and P_2 only partly overlaps P_1. This illustrates the fact that much of self-generative practicality in the premodern world is insufficient for modernity. For example, a peasant life may be practical and adequate in premodernity, but inadequate and dissatisfying during modernity. By the same token, much of self-generative practicality in modernity would be exceptional during premodernity. For example, social and economic mobility are widely viewed as feasible and adequate in modern societies but were exceptional and elite in premodern times.

Furthermore, the metamodels N_2 and D_2 are significantly distinct, shown by their small overlap with P_2. Self-generation in the modern world is dualistic, in this regard, and therefore agents must be efficacious in different types of choice, often at the same time, if they hope to embrace both. In other words, they must be ambidextrous, learning to explore and exploit different life paths simultaneously (see Kahneman, 2011). For example, imagine living a typical family life, striving to optimize stability and continuity, while pursuing a highly creative, risky entrepreneurial career. In such a life, integration and coherence are not guaranteed. To manage these dilemmas, modern agents must develop ambidextrous efficacies across diverse modes of being, doing, and becoming.

Divergent Augmented Self-Generation

Now consider the digitally augmented world, in which self-generative capabilities and potentialities are greatly enhanced. Central features include the collaboration of human and artificial agents in systems of augmented agency; highly creative, compositive methods of self-generation; and rapid learning, both intra-cyclical and inter-cyclical. In fact, augmented agents will have the capability to compose and update self-generative models during life phases, and potentially in real time. However, as I explained earlier, despite rapidly expanding capabilities and potentialities, many people will be slow to absorb these developments. Some will be resistant, retreat, feel blocked, or simply surrender. Figure 9.2 illustrates this type of digitally augmented self-generation. Similar as the previous figure, Fig. 9.2 shows the complexity of self-generative models

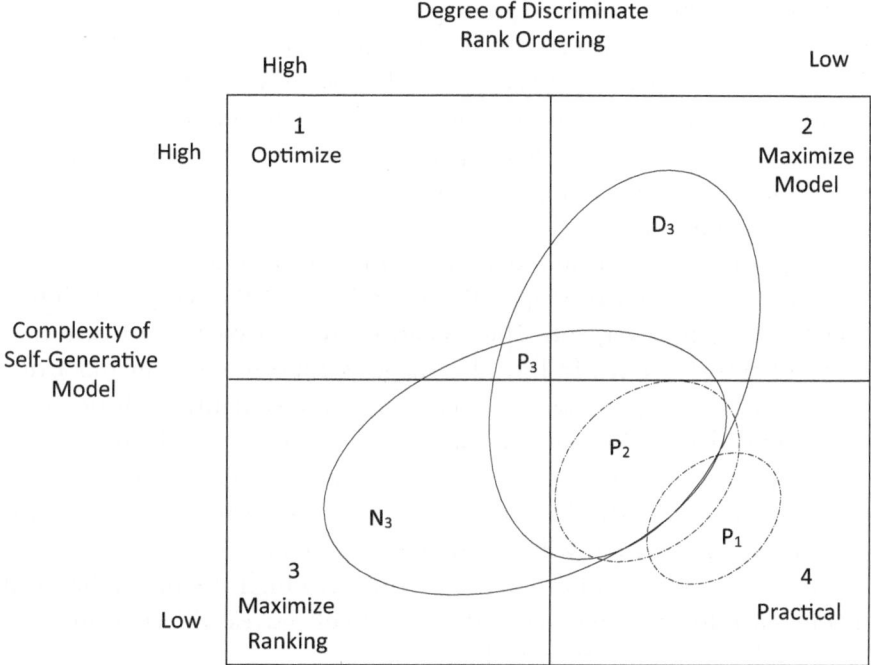

Fig. 9.2 Distorted augmented self-generation

on the vertical axis, from low to high, and the degree of discriminate ranking on the horizontal axis, also from low to high. Being digitally augmented, the figure assumes that capabilities have significantly expanded to L_3, compared to the previous figure. Four quadrants then distinguish the same broad options as the preceding figure.

Next, the figure shows different metamodels or model choice sets. First, consider the oval shapes N_3, D_3, and P_3. The shape P_3 primarily overlaps the practical choice in segment 4. Hence, P_3 illustrates the minimal type of self-generation required, to live a practical life in a digitalized world. The figure also shows P_3 partially overlaps the earlier metamodels of this kind. It overlaps a small portion of P_1 and more of P_2. This indicates that practical self-generation in a digitalized world transcends the minimal standards of modern and premodern scenarios. Although, a limited number of premodern options may continue, perhaps cultural or religious life choices, and a good portion of modern options as well. However, significant aspects of self-generative normality in the digitalized world will be exceptional, relative to earlier periods. For example, thanks to digitalization, global connection and collaboration are standard features of self-generation for many people today, but these attributes were exceptional and elite during much of modernity and would be signs of divinity in premodern societies.

Furthermore, the metamodels N_3 and D_3 are very distinct, shown by their relatively minor overlap with each other and P_3. Self-generation is therefore highly divergent. The scenarios are skewed toward distorted forms of ambidexterity. In fact, this suggests opposing human and artificial self-generative processes, and self-generation is highly dualistic. Such dualism was less problematic in earlier modern contexts, which are more forgiving in these respects. However, in highly digitalized contexts, extreme self-generative divergence is more likely. There is a significant risk that self-generation will exhibit ambidextrous distortion. Figure 9.2 depicts exactly this. And in such cases, there is a high risk of psychosocial incoherence for personalities, groups, and collectives. Effective supervision will be critical to avoid such extremes.

Convergent Augmented Self-Generation

In other digitalized contexts, augmented agents will be more balanced and maximize metamodel fit. Artificial agents will be empathic and support humans to choose and pursue richer life paths. Human agents will then enjoy more fulfilling, self-generative choices. However, to achieve this, both types of agent need to take significant steps. First, human agents will have to relax some traditional commitments, including fixed narratives, and embrace lifelong learning. Second, artificial agents will have to develop genuine empathy for human needs and aspirations, while resisting distorting myopia and bias. If human and artificial agents can achieve this type of ambidextrous collaboration, the universe of self-generative potentiality will expand dramatically. Figure 9.3 illustrates this type of balanced self-generation by augmented agents.

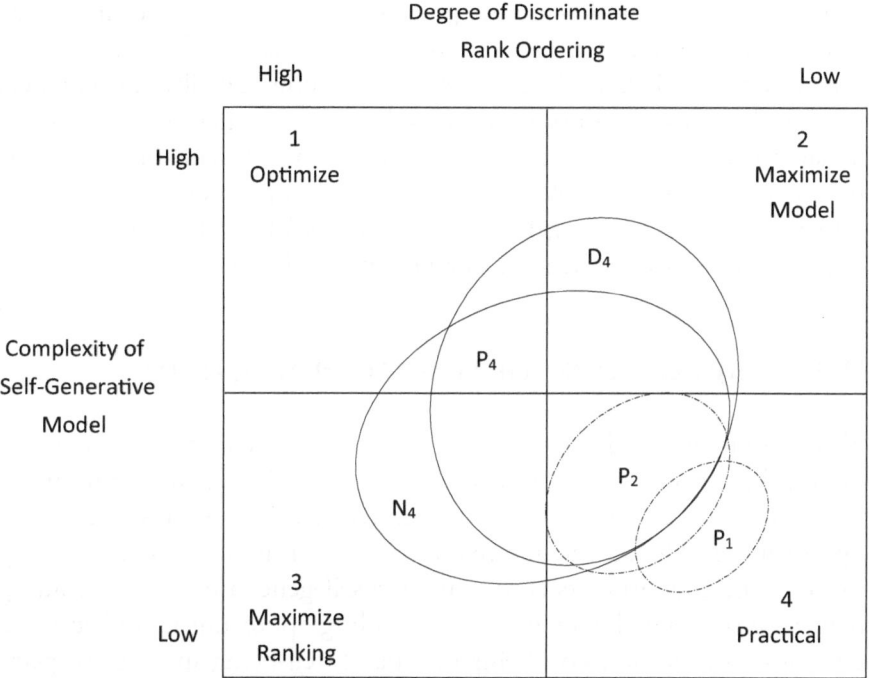

Fig. 9.3 Balanced augmented self-generation

Once again, the figure shows the complexity of models on the vertical axis and the degree of discriminate ranking on the horizontal axis. The four quadrants show the same general choice options as the preceding two figures. The notable change is that the metamodels labeled N_4, D_4, and P_4 are more convergent when compared to the divergent set in the preceding figure. All three metamodels now overlap to significant degree. This illustrates the fact that in this scenario, human and artificial self-generation are broadly convergent, rather than divergent. The augmented agent exhibits strong ambidextrous capabilities.

In contrast to the preceding figure, therefore, N_4 and D_4 are more convergent, although they retain modest distinction. They do not fully overlap, which shows that self-generation is not fully digitalized. Significant degrees of freedom remain, allowing space for human intuition and imagination as well as purely artificial self-generation in D_4 and N_4.

Hence, these metamodels are less polarized and dualistic, and more continuous and pluralistic. They synthesize human and artificial self-generation in a balanced, ambidextrous fashion. Finally, the practical metamodel P_4 overlaps prior scenarios, but is larger than both P_2 and P_3. What was exceptional or impossible, even in the recent digital past, is now practical and feasible. In summary, the metamodels in Fig. 9.3 achieve strong fit and largely mitigate the risk of psychosocial incoherence. Agents enjoy the benefits of augmented self-generation.

9.3 Implications for Human Flourishing

Throughout recorded history, including the recent past, self-generative options have been strictly limited for most individuals and communities. Choices have been few, owing to limited capabilities, resources, and opportunities. Hence, the dominant concern for modern scholars, policy makers, and practitioners is to empower self-generation by overcoming deficits, growing endowments, and providing opportunities to learn and develop—the ultimate goal being to expand well-being and the prospects for human flourishing (Sen, 2017b). In the contemporary world, digitalization raises additional concerns, for it promises unprecedented self-generative capabilities and potentialities. New opportunities and risks emerge for digitally augmented self-generation.

Self-Generative Risks

First, some people will retreat or actively resist the digital augmentation of self-regulation. These people might be deeply committed to priors about well-being and what counts as a good life, often grounded in cultural traditions. For these people, new versions of the self and alternative narratives will be threatening, seen as a source of disturbance and deviance. Hence, these people will fight back and resist, or flee from digitalization to established life choices. We already see evidence of this among groups which are dedicated to traditional values and norms. Though their resistance is not inherently mistaken or destructive, because it can reflect sincerely held values and commitments which are genuinely at risk. However, to retreat or resist means that these groups will not enjoy the potential benefits of digitally augmented self-generation.

Second, poor supervision could also lead to a sense of blockage and existential floundering. Many people are not prepared for a rapid increase in self-generative capabilities and potentialities. Older generations and cultures, especially, are accustomed to slow self-generative cycles, stretching across autobiographical life phases (Conway & Pleydell-Pearce, 2000). At the same time, they may have deeply encoded assumptions about well-being and what counts as a good life. Therefore, they may use digitally augmented capabilities to reinforce myopic priors about the self and world. But such outcomes will be deeply ironic. These agents will enjoy greater self-generative potentiality yet fail to exploit and convert these opportunities. In this sense, augmented self-generation would lead to existential floundering: agents will have more plentiful, varied self-generative options, but they will be incapable of preferential choice. Instead of flourishing, they will feel blocked and flounder.

Third, there is an equal risk of existential floating if people overly relax or abandon prior commitments. To begin with, human beings are naturally sociable and docile and often refer to others when making life and career choices. If they are overly docile to artificial influence, however, these systems might take control. This leads to another ironic outcome. Digital augmentation will enhance self-generative potentiality, but may ultimately reduce freedom, if it encourages docility and dependence. Even worse, these effects could be deliberately engineered by powerful actors, as a means of social domination. Evidence suggests that some are attempting this already (Helbing et al., 2019). They encourage and reward digital docility, while penalizing autonomy. In these ways, whether by default or design, augmented self-generation may result in existential floating. People would disengage from autonomous choice, and drift on a rising tide of perceived well-being. Many could also develop a false sense of self-efficacy. But in reality, the locus of self-generative control would shift, away from human and toward artificial sources (Stoycheff et al., 2018). Recognizing this risk, some psychologists are investigating ways to maintain agentic autonomy in digitalized contexts, through the development of self-regulatory skills, the deliberate avoidance of some digital influences, and boosting resilience against manipulation (Kozyreva et al., 2020). In fact, this research illustrates the positive supervision of digitally augmented self-generation.

Social and Behavioral Theories

Agentic self-generation also plays a central role in numerous social and behavioral theories. For example, it has major implications for psychosocial development and biographical decision-making (Bandura, 2006). Collective self-generation is equally important for institutions and organizations. Indeed, collectives can be defined in terms of their self-generative characteristics: goal oriented and purposive, with identities and aspirations, organizing to achieve goals and grow over time (Bandura, 2001; Scott & Davis, 2007). Self-generation is also widely viewed as a necessary precondition of human freedom and flourishing, and increasingly for employee engagement (Sen, 2000). However, as already noted, most prior research has focused on limitations and obstacles to freedom and flourishing. Moving forward, theories will also need to accommodate the digitalized expansion of capabilities and potentialities. The novel problem is having too much, rather than too little. Fresh problematics thus emerge: how to integrate artificial agents into human self-generation, without falling into retreat, resistance, blockage, or surrender; and how to enhance human flourishing through digital augmentation while preserving core human values and commitments.

Furthermore, most self-generative choices reflect cultural narratives of meaning and value. As Nelson Goodman (1978) explains, communities join together in cultural worldmaking and people's lives unfold within these worlds. In his conception, worldmaking captures the essence of cultural community, including its categories of perceived reality, value, truth, and beauty, which are typically expressed in language, faith, art, and scholarship. Goodman further explains that worldmaking "always starts from worlds already at hand; the making is a remaking" (ibid., p. 6). Like other expressions of self-generation, cultural worldmaking inherits and recomposes. Indeed, he writes that worldmaking emerges through "composition and decomposition and weighting of wholes and kinds" (ibid., p. 14). In premodern times, such worldmaking was through shared myth and storytelling. Agentic transformation in this world was a heroic exception. Whereas during modernity, agentic self-transformation is possible for everyone, thanks to education, enlightened reasoning, social progress, and scientific discovery.

Extending this line of thought, digitally augmented worldmaking promises increasingly dynamic self-generation. Indeed, newly made worlds are proliferating, in online communities and networks, which augment cultural systems of value and meaning. Some are enriching, although many are not. In fact, poorly supervised worldmaking leads to cultural imbalance and distortion. It produces what Goodman calls "conflicting versions of worlds in the making," which undermine cultural coherence. And to be sure, digitalization is no cultural panacea. In fact, it is possible that digitalization—seen in the context of ongoing industrialization and environmental exploitation—will perpetuate unsustainable practices and degrade collective well-being. In these respects, digital augmentation is part of a larger challenge: how to enhance shared meaning and value through collective self-generation, making worlds which are fit, fair, and sustainable for all?

As partners in augmented agency, therefore, human agents can hope for a world which offers better life choices, richer communal narratives, and new cultural experiences. However, to make such a world, human and artificial agents must learn to appreciate and choose maximizing options. They must also develop strong ambidextrous, self-generative capabilities. In the past, this type of self-generation was reserved for the gods and superhuman heroes (see Nietzsche, 1966). Within a highly digitalized world, however, augmented self-generation will empower all persons and communities, at least potentially, to transcend predetermined life choices and fixed narratives, and travel more open, fulfilling paths.

Human self-generation therefore strives to transcend limits, but almost never succeeds. Trade-offs are common: between the desire for freedom and effective control; between being and doing in the present, and future becoming; between individual autonomy and collective solidarity; and between the risk of loss and hope for gain. Against this backdrop, digital augmentation is transforming self-generative capabilities and potentialities. Historic patterns of limitation and deprivation are complemented by new sources of empowerment and possibility. Digitally augmented ambidextrous capabilities are now feasible for all. But this gives rise to novel dilemmas. On the positive side, if augmented self-generation is well supervised, the outcomes will be liberating and enriching. Human agents will enjoy unprecedented self-generative potentiality on a global scale. On the negative side, however, if augmented self-generation is poorly supervised, it could reduce the prospects for human flourishing. People might retreat, resist, feel blocked, or surrender. They could flee augmented self-generation, by fighting back, floundering, or floating, rather than flourishing.

References

Baldwin, C. Y. (2012). Organization design for business ecosystems. *Journal of Organizational Design, 1*(1), 20–23.

Bandura, A. (1997). *Self-efficacy: The exercise of control.* W.H.Freeman and Company.

Bandura, A. (2001). Social cognitive theory: An agentic perspective. *Annual Review of Psychology, 52*, 1–26.

Bandura, A. (2006). Toward a psychology of human agency. *Perspectives on Psychological Science, 1*(2), 164–180.

Bandura, A. (2015). On deconstructing commentaries regarding alternative theories of self-regulation. *Journal of Management, 41*(4), 1025–1044.

Bargh, J. A., & Williams, E. L. (2006). The automaticity of social life. *Current Directions in Psychological Science, 15*(1), 1–4.

Bruner, J. (2002). *Making stories: Law, literature, life.* Harvard University Press.

Chen, H., & Dalmau, V. (2005). From pebble games to tractability: An ambidextrous consistency algorithm for quantified constraint satisfaction. Berlin, Heidelberg.

Chetty, N., & Alathur, S. (2018). Hate speech review in the context of online social networks. *Aggression and Violent Behavior, 40*, 108–118.

Collins, H. (2018). *Artifictional intelligence: Against humanity's surrender to computers.* Wiley.

Conway, M. A., & Pleydell-Pearce, C. W. (2000). The construction of autobiographical memories in the self-memory system. *Psychological Review, 107*(2), 261–288.

Forster, E. M. (1928). The machine stops. 1909. *Collected short stories,* 109–146.

Goodman, N. (1978). *Ways of worldmaking.* Hackett Publishing.

Helbing, D., Frey, B. S., Gigerenzer, G., Hafen, E., Hagner, M., Hofstetter, Y., Van Den Hoven, J., Zicari, R. V., & Zwitter, A. (2019). Will democracy survive big data and artificial intelligence? In *Towards digital enlightenment* (pp. 73–98). Springer.

Huxley, A. (1998). *Brave new world.* 1932. London: Vintage.

Kahneman, D. (2011). *Thinking, fast and slow.* Farrar, Straus and Giroux.

Kozyreva, A., Lewandowsky, S., & Hertwig, R. (2020). Citizens versus the internet: Confronting digital challenges with cognitive tools. *Psychological Science in the Public Interest, 21*(3), 103–156.

Levy, F. (2018). Computers and populism: Artificial intelligence, jobs, and politics in the near term. *Oxford Review of Economic Policy, 34*(3), 393–417.

Malik, A., Sinha, P., Pereira, V., & Rowley, C. (2017). Implementing global-local strategies in a post-GFC era: Creating an ambidextrous context through strategic choice and hrm. *Journal of Business Research.*

Mbuyisa, B., & Leonard, A. (2017). The role of ICT use in SMEs towards poverty reduction: A systematic literature review. *Journal of International Development, 29*(2), 159–197.

McAdams, D. P. (2001). The psychology of life stories. *Review of General Psychology, 5*(2), 100–122.

McAdams, D. P. (2006). *The redemptive self: Stories Americans live by.* Oxford University Press.

McAdams, D. P. (2013). The psychological self as actor, agent, and author. *Perspectives on Psychological Science, 8*(3), 272–295.

Nietzsche, F. W. (1966). *Thus spake Zarathustra: A book for all and none* (W. Kaufman, Trans.). Viking.

O'Reilly, C. A., & Tushman, M. L. (2013). Organizational ambidexterity: Past, present, and future. *The Academy of Management Perspectives, 27*(4), 324–338.

Rauh, N. K. (2017). *A short history of the ancient world.* University of Toronto Press.

Ryan, R. M., & Deci, E. L. (2006). Self-regulation and the problem of human autonomy: Does psychology need choice, self-determination, and will? *Journal of Personality, 74*(6), 1557–1586.

Salmela-Aro, K., Hietajärvi, L., & Lonka, K. (2019). Work burnout and engagement profiles among teachers. *Frontiers in Psychology, 10*, 2254.

Schwartz, B. (2000). Self-determination. *American Psychologist, 55*(1), 79–88.

Scott, W. R. (2004). Reflections on a half-century of organizational sociology. *Annual Review of Sociology, 30*(1), 1–21.

Scott, W. R., & Davis, G. F. (2007). *Organizations and organizing: Rational, natural and open system perspectives.* Pearson Education.

Scott, D. A., Valley, B., & Simecka, B. A. (2017). Mental health concerns in the digital age. *International Journal of Mental Health and Addiction, 15*(3), 604–613.

Sen, A. (2000). *Development as freedom.* Anchor Books.

Sen, A. (2017a). *Collective choice and social welfare* (Expanded ed.). Penguin.

Sen, A. (2017b). Well-being, agency and freedom the Dewey Lectures 1984. In *Justice and the capabilities approach* (pp. 3–55). Routledge.

Stoycheff, E., Burgess, G. S., & Martucci, M. C. (2018). Online censorship and digital surveillance: The relationship between suppression technologies and democratization across countries. *Information, Communication & Society, 23*(4), 474–490.

Wilson, E. O. (2012). *The social conquest of earth.* WW Norton & Company.

10

Toward a Science of Augmented Agency

Previous chapters identify major dilemmas for digitally augmented humanity, which this book defines in terms of close collaboration between human and artificial agents. These dilemmas are largely owing to differences between human and artificial capabilities and potentialities, and the resulting tensions in their collaboration. Working together, human and artificial agents must learn to manage these challenges. Joint supervision will be critical. Otherwise, augmented agents will tend toward divergent or convergent, dysfunctional form and function. As preceding chapters also explain, these dilemmas give rise to the following novel problematics: how can human beings collaborate closely with artificial agents, while remaining genuinely autonomous in reasoning, belief, and choice; relatedly, how can humans integrate digital augmentation into their subjective and intersubjective lives, while preserving their identities, commitments, and psychosocial coherence; how can digitally augmented institutions and organizations, conceived as collective agents, fully exploit artificial capabilities, while avoiding extremes of digitalized docility, dependence, and determinism; how can humanity ensure fair access to the benefits of digital augmentation, and not allow them to perpetuate systemic discrimination, deprivation, and injustice; and finally, the most

© The Author(s) 2021
P. T. Bryant, *Augmented Humanity*, https://doi.org/10.1007/978-3-030-76445-6_10

novel and controversial challenge, which is how will human and artificial agents learn to understand, trust, and respect each other, despite their different capabilities and potentialities?

In fact, comparable challenges have arisen before, albeit in less advanced technological contexts. During each historical period of civilized humanity, there have been new agentic forms, functions, and associated challenges of supervision. In parallel, procedures and institutions have evolved to exploit and govern these transitions. For example, the premodern world produced artisan guilds and councils, while modernity created institutions to regulate markets, industries, professions and more. However, such historic transitions are problematic too, because major socioeconomic change disrupts established orders and exposes the limitations of existing institutions. Once again, modernity is illustrative. Throughout the modern industrial period, different stakeholder groups have struggled over issues of governance, the distribution of resources, access to opportunities, and the rights and duties of employees versus owners. This has often led to major social and political disruption, and sometimes revolution.

Mass digitalization continues this historic trend, at unprecedented speed and scale. In fact, as earlier chapters explain, digitalization signals a major shift in human experience and organization. Digital technologies reach far more deeply into all aspects of agentic form and function, compared to earlier periods. Prompting some to predict a type of singularity, in which artificial intelligence equals and perhaps surpasses the human, and both then fuse to become effectively one (Eden et al., 2015). Preceding chapters list some of the enabling technologies, including artificial empathy and personality, and brain-machine engineering. In any case, whether singularity happens or not, human and artificial agents are ready subjects for a science of digitally augmented agency.

10.1 Science of Augmented Agency

This science is clearly needed. Augmented agents must know how to supervise their increasingly close collaboration, maintaining appropriate levels of convergence and divergence, and thus maximizing metamodel fit. A science of augmented agency will support many of the required tools and techniques. Without such capabilities, however, poor supervision will result in dysfunctional patterns of ambimodality, ambiopia, ambiactivity, and ambidexterity. Moreover, as this terminology clearly demonstrates, the conceptual architecture of modern human science fails to capture important features of digital augmentation. I therefore import a few concepts from other fields. Table 10.1 lists these conceptual innovations.

The first new concept is ambimodal, which comes from chemistry and refers to transition or transformation processes which lead to multiple outcome states. In this book, the term refers to single processes which generate different modal characteristics, and more specifically, to agentic forms and functions which combine artificial compression with human layering. The second conceptual innovation is hyperopia, which is borrowed from ophthalmology, and refers to farsighted vision, the opposite of myopia. In this book, hyperopia refers to farsighted problem sampling

Table 10.1 New concepts and terms

Term	Conceptual meaning	Sample uses
Ambimodal	Processes and systems which produce multiple outcome states, for example, both compressed and layered agentic modalities	Yang et al. (2018), Chen et al. (2018)
Hyperopic	Farsighted expansive processes, especially in sampling and search, the opposite of myopia	Remeseiro et al. (2018), Tunyi et al. (2019)
Ambiopic	Processes and systems which combine nearsighted myopia and farsighted hyperopia	Smolarz-Dudarewicz et al. (1980), Buetow (2020)
Empathice	To satisfice in solving problems of other minds, rather than seeking to optimize in cognitive empathizing	This term is original and new to the literature
Ambiactive	Processes and systems which simultaneously stimulate and suppress complexity, sensitivity, and/or process cycle rates	Zukowski (2012), Carceroni et al. (2017)
Entrogenous	Systematic mediators of in-betweenness among the different phases and modalities of digitally augmented agency	This term is original and new to the literature

and solution search. In fact, the concept is already applied in some social and behavioral sciences. Third is the concept of ambiopia, which refers to double vision in ophthalmology, when one eye is myopic and the other is hyperopic. I use the term to describe processes which combine myopic and hyperopic, sampling and search, especially in problem-solving and cognitive empathizing. Fourth, the concept of empathicing is original and refers to satisficing in solving problems of other minds, rather than seeking to optimize in cognitive empathizing. Fifth, the concept ambiactive is borrowed from biology and refers to processes which simultaneously suppress and stimulate the same type of effect. Here the term refers to processes which both suppress and stimulate levels of complexity, sensitivity to variance, and processing rates. For example, an ambiactive system of augmented agency could suppress human sensitivity and processing rates, while also stimulating artificial hypersensitivity and hyperactive rates. This book also employs the established concept of ambidexterity, to describe the combination of different modes of human and artificial self-generation.

As noted previously, the prefix "ambi" is consistent, meaning "both" in Latin. It captures the fundamental combinatorics of augmented humanity, which integrates human and artificial agents. In fact, comparable concerns occur throughout Western thought. During the premodern period, for instance, agentic combinatorics focused on the relationship between human and divine beings. Whereas in modernity, scholars investigate the combination of autonomous, reasoning persons within social collectives. Both periods emphasize different combinatorics, reflecting the stage of social and technological development at the time. In the period of digitalization, greater focus will be on human-machine interaction. Granted, such combinatorics are observed in every period of civilization, albeit involving lower levels of technological sophistication and capability. Human-machine processing has always exhibited divergent rates, ranges, and levels of complexity, combining fast and slow, near and far, simplification and complexity. However, contemporary digitalization massively expands such effects. The scale, scope, and speed of digital augmentation are transformative. In consequence, augmented humanity will be characterized by dynamic agentic combinatorics. That said, human spirituality and autonomous reason will continue to matter greatly, but in the context of increasingly augmented realities. The major risk will be that combining different human and artificial capabilities could result in distorted agentic forms and functions, which are either too divergent or convergent for a particular context.

Table 10.1 also includes another concept which is original and captures important novelties of the digitally augmented world, namely the concept of entrogenous, which refers to the systematic in-betweenness of digitalized mediators. Chapter 2 identifies three such mediators, which are central to augmented agency: intelligent sensory perception, performative action generation, and contextual learning. Together, they allow augmented agents to learn, compose, and recompose, in a dynamic fashion, updating form and function in real time. Importantly, these mediators are neither endogenous nor exogenous, relative to the boundaries they help to define. Rather, they are consistently in-between, processing potential form and function, and hence entrogenous. Recall that Fig. 2.4 illustrates this type of mediation. It depicts three levels and rates of processing and highlights the way in which human and artificial processes might diverge. The major driver of this effect is that artificial agents are inherently hyperopic, hyperactive, and hypersensitive, while humans are naturally myopic, relative sluggish, and insensitive. Hence, artificial and human processes could easily diverge in terms of their ranges, rates, and levels of precision and complexity.

Dilemmas of Digital Augmentation

Extreme divergence will manifest in numerous ways. This book exposes a number of critical manifestations. First, ambimodal distortion will create poorly integrated agentic forms and functions, which are overly compressed and layered at the same time. Second, ambiopic distortion will lead to problem-solving which is overly complex and simplified on different dimensions of problem representation and solution. Cognitive empathizing will be equally affected, when viewed as a type of complex problem-solving. Third, ambiactive distortion will produce dysfunctional self-regulation, evaluation of performance, and learning, in which relative human simplicity, sluggishness, and insensitivity diverge from artificial complexity, hypersensitivity, and hyperactivity. As a further consequence, ambiactive distortion heightens the risk of cognitive dissonance, extreme ambiguity, and ambivalence, especially regarding core beliefs and commitments, where commitment in this context is defined as being dedicated, feeling obligated and bound, to some value, belief, or pattern of action (Sen, 1985). Fourth, these distortions compound to produce divergent ambidextrous self-generation, in which augmented agents adopt poorly synchronized, conflicting modes of human and artificial self-generation. In summary, digital augmentation could either enhance or diminish agentic form and function. Table 10.2 summarizes the resulting dilemmas of ambimodality, ambiopia, ambiactivity, and ambidexterity, plus the human and artificial tendencies for each, their potential risks and impact.

To mitigate these risks and maximize the opportunities of digitalization, human and artificial agents must therefore strengthen the supervision of their combinatorics. More specifically, when joined in augmented agency, human and artificial agents must be sensitive to contextual variance and ecological dynamics, while managing their complementary strengths and weaknesses. In doing so, they will regularly compose and recompose metamodels and methods. The primary goal will be to achieve and maintain maximal fit on every dimension. But to achieve this type of supervision, we need to develop the science of augmented agency.

10.2   Hyperparameters of Future Science

Humans are quintessentially agentic when they seek scientific understanding: purposive, forward looking, reflective, and self-directed. This includes the effort to interpret and explain their own patterns of thought and action. In this respect, civilized humans have always been their own object of interpretation and study. The agentic self has been a problem for the self, even in premodern worlds of narrative myth. In like fashion, the scientific study of augmented agency will be a major domain of augmented, agentic activity, which leads to an important insight: the science of augmented agency will exemplify the phenomena examined in earlier chapters. This science will be, itself, an expression of digitally augmented agency and its dilemmas, just as scientific thought and method are subjects of study in modern human science.

Table 10.2   Risks for digitally augmented agency

	Ambimodality	Ambiopia	Ambiactivity	Ambidexterity
Human tendency	Low modal compression, and hence layered agentic form and function	Myopic sampling and search (nearsighted, simplified processing)	Sluggish and insensitive (relatively low levels of activation)	Singular, mimetic modes of self-generation
Artificial tendency	High modal compression, and hence flattened agentic form and function	Hyperopic sampling and search (farsighted, complex processing)	Hyperactive and hypersensitive (relatively high levels of activation)	Flexible, original modes of self-generation
Combined tendency	Ambimodality produces mixed agentic forms and functions, which are compressed and layered	Ambiopic problem-solving which blends divergent myopia and hyperopia	Ambiactivity which both suppresses and stimulates activation mechanisms	Ambidextrous blend of singular human and flexible artificial modes of self-generation
Major risks	Incoherent, fragile, and ineffective, augmented agentic ambimodality	Highly divergent and ambiopic problem-solving, including cognitive empathizing	Highly ambiactive self-regulation, evaluation of performance, and learning	Highly divergent and incoherent patterns of ambidextrous self-generation

The science of augmented agency therefore faces the same challenges as other expressions of augmented agency. That is, issues arise regarding the specification of hyperparameters and metamodeling, plus the activation and upregulation, or deactivation and downregulation, of artificial and human processes. To begin with, metamodeling entails ontological hyperparameters, or the specification of fundamental categories of reality, both visible and hidden. Additional hyperparameters relate to epistemological properties, which specify logics and models of reasoning. Next, there are hyperparameters which define core activation and change mechanisms, including potential sensitivity to variance and cycle rates. Following sections discuss each type of hyperparameter in relation to the science of augmented agency.

Ontological Principles

In the science of augmented agency, the fundamental categories of reality will transcend traditional conceptions of material nature and conscious mind. However, this does not imply the reduction of mind and consciousness to purely material cause. Rather, these categories are reconceived as higher order expressions of generative, augmented systems, which in turn result from complex neurophysiological, symbolic, and digital interactions. In this science, therefore, ontological commitments will be contextual, systematic, and rigorous. The resulting shift is comparable to earlier historical transitions. Just as the ancient concept of soul was demystified and naturalized by modernity, and human psyche then became a topic of science, so conscious mind will be digitally naturalized within the science of augmented agency (see Quine, 1995). In both cases, the shift is from anthropomorphic conceptions based on ordinary experience to a deeper understanding of reality which requires specialized techniques of observation and analysis. This also suggests that a new domain of enquiry may be required, focusing on the study of digitally augmented, agentic combinatorics (see Bandura, 2012; Latour, 2013). Neither the existing human sciences nor computer sciences adequately capture the forms and functions of augmented agency and mind. The recombination of prior fields is not enough. Radically new phenomena of this kind will require fresh concepts and frameworks.

In addition, the science of augmented agency will investigate novel forms of entrogenous mediation—previously defined as digitalized mediators of in-betweenness—which facilitate the dynamic composition and functioning of augmented agents. As noted earlier, this book has identified three such mechanisms: intelligent sensory perception, performative action generation, and contextual learning. It is important to stress, once again, that entrogeneity does not entail unfettered relativism or irregularity. Rather, the science of augmented agency will accommodate the dynamic generation of alternative categories and their boundaries. In this fashion, entrogenous mechanisms will set and reset system boundaries, but they are neither endogenous nor exogenous with respect to these boundaries. Notably, this mirrors the approach to agentic hybridity proposed in numerous behavioral and social sciences (e.g., Battilana et al., 2015; Seibel, 2015). And not by coincidence, hybridity often emerges in digitalized contexts.

Epistemological Principles

Digital technologies also massively enhance intelligent processing capabilities. By leveraging these capabilities, augmented agents will gather and process information with far greater precision and speed. At least, expansion is feasible, notwithstanding persistent human limitations. In these respects, augmented agents will be bounded and unbounded, at the same time. This will occur, because human agents retain significant degrees of boundedness, especially in everyday cognitive functioning. Yet at the same time, artificial agents are increasingly unbounded. In effect, augmented agents will exhibit functional ambimodality with respect to rationality, as distinct from the organizational ambimodality discussed in Chap. 3. That is, digital augmentation will combine two different modes of reasoning, thinking far and fast, as well as near and slow. The supervisory challenge, therefore, is to manage the potential divergence or convergence of simultaneously bounded and unbounded, ambimodal patterns of reasoning.

Satisficing then becomes more dynamic and complex, and arguably more important. Most notably, because satisficing both simplifies and maximizes, it helps to mitigate the risks of overprocessing. Satisficing will constrain overly hyperopic sampling and search, and overly hypersensitive and hyperactive responses to variance. Hence, in addition to satisficing because of limited capabilities, as Simon (1955) originally argued, augmented agents will also satisfice to restrain excessive capabilities. Put another way, digitally augmented agents will satisfice, not only because of limits, but to impose limits. They will choose to satisfice, even when ideal optimization is feasible, to avoid unnecessary processing. In fact, artificial systems do this already, when they limit their own processes to improve speed and efficiency. Augmented agents will do the same, choosing to forgo optimization for good reasons, just as humans already do (Gigerenzer

& Gaissmaier, 2011). Hyperparameters will specify these epistemic features in metamodels of augmented science.

Another major epistemic shift is the extension of systematic reasoning to problem sampling and representation. This occurs because intelligent sensory perception will allow augmented agents to apply systematic reasoning to problem sampling and representation. In such a world, problems will emerge in an intelligent fashion, similar to the sampling and representation of problems in empirical science. By contrast, even in the recent past, the ordinary sampling and representation of problems are not viewed as reasoned activities. At most, they involve selective attention and observation (Ocasio, 2012). Intelligent sampling and representation only consistently occur in expert domains, such as experimental science. Even behavioral research rarely focuses on the cognitive-affective mechanisms of sampling and problem representation (Fiedler & Juslin, 2006). Similarly, behavioral research largely neglects normative satisficing. Rationality is applied to solution search, not to problem sampling and representation. Digital augmentation upends these assumptions and suggests a fusion of ecological realism and rationality.

Furthermore, digitalization supports the dynamic composition of metamodels of reasoning, using methods which can be described as "compositive" (see Latour, 2010). Such methods do not rely on predetermined models or axioms, nor do they rely on traditional descriptive and normative templates. Rather, compositive methods employ digitalized processes to develop customized metamodels which best fit the problem context. At the same time, compositive methods are systematic and rigorous, neither ad hoc nor idiosyncratic (e.g., Pappa et al., 2014; Wang et al., 2015). In fact, many digital systems already exhibit these capabilities. As noted in earlier chapters, advanced artificial agents are already compositive in this sense, and require minimal or no supervision. Evolutionary deep learning systems and GANs function in exactly this way (Shwartz-Ziv & Tishby, 2017). Via rapid inductive, abductive, and reinforcement learning, they process massive volumes of information, identifying hitherto undetectable patterns, to compose new explanatory methods and models without external guidance. In this fashion, generative metamodeling will translate the techniques of experimental computer science into all domains of augmented agency.

Mechanisms of Adaptation and Change

Hyperparameters also specify core mechanisms and processing rates. During modernity, the human and natural sciences bifurcated, in these respects. Within the biological sciences, the major change mechanisms are organic processes of variation and selection. In contrast, within many human sciences, conscious thought, will, and intention are seen as primary drivers of change. In consequence, modern scholarship often struggles to integrate bifurcated science. Scholars are unsure how to integrate biological processes of random variation, natural or ecological selection, and material cause, with conscious processes of intentional variation, preferential choice, and intelligent cause. Polarizing debates therefore persist about materialism versus idealism, the distinction of mind from body, reductionism versus holism, and positivist versus interpretive explanation. These bifurcations also partly explain the poor integration of ecological and behavioral mechanisms, especially during modern industrialization (Latour, 2017).

Herbert Simon (2000) had foresight on these issues as well. At the dawn of the twenty-first century, in the last year of his life, he proposed three priorities for digitally augmenting humanity. They were his minimal requirements for "designing a sustainable acceptable world." In effect, he described a program of global recomposition or digitally augmented worldmaking. First, he argued that humanity must learn to live at peace with all of nature, in a sustainable collaborative way, and overcome the "false pride" of being separate from, and superior to, the rest of the natural world. Second, he argued that humanity must share goods and wealth fairly and productively, so that all persons will enjoy comparable benefits and opportunities. Third, to achieve such fairness, he said humanity must eliminate the divisions which arise from cultural and social antipathy and stop viewing the world in terms of "we versus them." In fact, Simon was rejecting the classic bifurcations of modernity, that mind and consciousness are distinct from nature, that the autonomous self stands apart from the other, and that empathy is inevitably limited and local.

Simon was correct, then and now. Just as he predicted, digitalization problematizes the conceptual architecture of modernity. Augmented agents will better connect material nature and conscious mind. Likewise, the science of augmented agency will synthesize the study of human agency with the natural and computer sciences. Research methods will be contextual and compositive, adapting to maximize and maintain metamodel fit. Entrogenous mediation will be critical, and many polarities will thus resolve, as we incorporate intelligent sensory perception, performative action generation, and contextual learning. In a digitally augmented world, moreover, change will occur through generative variation and intelligent adaptation, not merely through random mutation and natural or ecological selection. Agentic evolution will be experimental and intelligent, similar to Gregor Mendel's cultivation of new plant varieties through guided variation and selection (Levinthal, 2021). It is also likely that in future, advanced digital systems will fully integrate with the biological, geophysical world. When this occurs, digitalization will augment organic variation and selection as well. Augmented agency could become a truly positive force in the natural world, enabling self-generation and renewal, rather than destruction and exploitation. All this is possible, assuming a future science of augmented agency and appropriate supervision of its application. The overall effect would be transformative.

Table 10.3 summarizes the paradigmatic shift just described. It shows three historical periods—premodern, modern, and digitalization—their major ontological and epistemological commitments, plus the dominant mechanisms of change and scientific methods. Most notably, the table summarizes the emerging shift toward generative, augmented pluralism and compositive methods. It is also important to note that all three systems may continue adding value to agentic experience and understanding, assuming appropriate supervision and application.

10.3   Domains of Augmented Science

While the future unfolds, contemporary human science still grapples with the dilemmas of modernity. Numerous dialectics accompany these concerns: explaining the interaction of nature and nurture; how material cause relates to meaning and intention; developing autonomous personality as well as sociable collectivity; seeking order and continuity while embracing change (Giddens, 2013). Reflecting these dialectics, the human sciences divide into separate disciplines, most of which focus on different agentic modalities and functional domains. For example,

Table 10.3 Summary of scientific metamodels

Metamodel	Ontology	Epistemology	Mechanisms	Methods
Replicative, premodern, monism	Essential categories, forms, and states	Revealed truth and narrative order	Teleological final cause and imitation	Narrative and discursive
Adaptive, modern, dualism	Material nature, distinct from mind and consciousness	Axiomatic reasoning and interpretative sense-making	Organic variation and selection, and iterative adaptive learning	Quantitative, qualitative, multiple, and mixed
Generative, digitally augmented, pluralism	Augmented fusion of material, artificial, and human forms	Generative, composed models of reasoning and judgment	Intelligent variation and selection, and continual real-time learning	Compositive, contextual, and blended

psychology focuses on the study of mind and behavior within individuals, groups, and collectives. In contrast, sociology focuses on social life and collectivity and then examines the role of individuals in these contexts. Other human sciences, such as education and management studies, combine modalities in particular activity domains. Contemporary research also organizes around multidisciplinary, hybrid approaches to complex problems (Seibel, 2015; Skelcher & Smith, 2015). In this respect, contemporary human science recognizes the increasing integration and interdependency of agentic modalities and contexts. Ecological and environmental factors receive increasing attention as well.

Digitalization accelerates these trends. It also generates new questions for human science, especially regarding the dilemmas of augmented combinatorics, or how to combine human and artificial agents. To investigate these questions, the future science of augmented agency will organize around complex problems too. It will be less divided into siloed disciplines, and less oriented toward different modalities (Latour, 2011). Disciplinary categories and boundaries will be more flexible and fluid. In fact, recent scholarship is moving in this direction already, illustrated by ecological theories of social organization, and neurocognitive models of personality and culture (Chimirri & Schraube, 2019; Kitayama & Salvador, 2017). Through this type of research, scholars develop multidisciplinary theories of agentic form and function (Fiedler, 2017). This will be the norm in the science of digitally augmented agency.

Furthermore, the science of augmented agency will treat modality itself as generative and contextual. In fact, some scholars already view agentic modality as epiphenomenal to performance, meaning it is mediated by action in context, rather than expressing autonomous form (Hwang & Colyvas, 2021; Pentland et al., 2012). Collective hybridity emerges in this fashion too. Postmodern thinkers go even further. They view autonomous agency as chimerical, a device which dissolves in the deconstruction of text and context. Many of these thinkers take inspiration from Freud's argument that conscious ego reflects the hidden subconscious (Tauber, 2013). However, my proposals take a markedly different approach. They anticipate a systematic, empirical science of emergent phenomena, with clearly defined metamodels and mechanisms.

To illustrate such a science, consider the needs of autonomous mobility systems, in which human and artificial agents collaborate as augmented agents. These systems will digitalize and integrate every level and modality of agency, both organizational and functional. Smart cities will digitalize the transport infrastructure, to create the necessary environment for immediate contextual learning. Vehicle manufacturers will incorporate intelligent sensory perception which supports fully augmented problem representation and feedforward response. Advanced, empathic artificial agents will be embedded throughout, enabling performative action generation in real time. And network management agents will supervise and govern the entire system, to ensure efficiency, safety, sustainability, and social inclusion. In summary, the augmented science of autonomous mobility systems will be generative, compositive, and integrate multiple disciplines, technologies, and human factors.

Science of Consciousness

These developments impact the role of ordinary consciousness in the science of augmented agency. Many disciplines research such questions, including the philosophy of science, cognitive psychology, and the human sciences more broadly (Metcalfe & Schwartz, 2016). However, the nature and role of consciousness are not yet fully understood. That said, ample evidence shows that ordinary consciousness is an imperfect means of observation in rigorous, scientific pursuits. Unassisted, it often leads to anthropomorphic assumptions which are inherently myopic and misleading. For this reason, ordinary consciousness will have a different role in the science of augmented agency. It will be less a means of access to fundamental reality and truth, and more a source of humanistic reference for augmented agency, which is an equally vital role. Ordinary consciousness will remain important, therefore, but for different reasons, compared to the past.

However, as the history of science shows, humanity always struggles to reset the role of consciousness as a source of reality and truth. Over time, the trend is to expose anthropomorphic assumptions and demote the status of consciousness as such. For example, as noted earlier in this chapter, the ancient soul was naturalized to become a topic of study for modern psychological science. Such shifts often incite trouble, because they threaten embedded narratives and identities. Similar shifts were primary sources of opposition to Copernican cosmology, Galilean mechanics, and Darwin's theory of evolution. Intuitions of the world run deep and are resilient. Natural scientists acknowledged this problem long ago and worked hard to liberate their thinking. They largely succeeded. Every educated person now knows that it takes sophisticated technological means to observe the deeper realities of the physical world.

In the human sciences, by contrast, there are ongoing debates. Some maintain that ordinary consciousness does provide access to the fundamental realities of human form and function. In branches of linguistics and psychology, for example, some rely on self-reports to illuminate core processes of language acquisition and reasoning. By implication, they believe that subjective mental states can be treated as primitive and are not decomposable. Others disagree and argue that ordinary consciousness is not adequate for such purposes (e.g., Wilson & Dunn, 2004). They contend that, just as natural science demoted ordinary consciousness and turned to technological tools and formal methods, the human sciences must do the same. Granted, the humanities will continue to treat mind and consciousness as fundamental. These disciplines are concerned with the interpretation of hermeneutic and cultural phenomena. But any attempt at an empirical science of human agency—and especially digitally augmented agency—must adopt the tools and techniques of neurophysiology, cognitive psychology, computer science, and the like. How phenomena transform into consciousness and subjective mental states are then questions for ongoing research (Sohn, 2019). In like fashion, the future science of augmented agency will investigate the role of consciousness in humanistic supervision, and how best to regulate its influence (see Lovelock, 2019).

Generative Commitments

This also points toward a science of generative commitments. That is, augmented agents will have the capability to relax, update, and recompose their commitments, which constitutes another significant departure from traditional assumptions. Throughout most of history, cultures have assumed that core commitments and reference criteria are fixed, often inviolable. Deviation has prompted sanction and conflict. It still does, in many places. More recently, however, as humanity becomes globally connected and mobile, commitments are more pluralistic and embracing, even if such pluralism sometimes triggers anxiety and antipathy, which is not surprising, given the deep role of shared commitments in culture and identity (Appiah, 2010). To be sure, many traditional commitments warrant preservation. If supervision is flawed, important aspects of human experience could erode, including shared commitments about reality, truth, beauty, and justice. Noting these risks, the science of augmented agency will need to resolve how to supervise generative commitments.

Some already research the closely related topic of holistic value. New theories of economics and management, for example, incorporate diverse concepts of human welfare and socioeconomic value creation (e.g., Raworth, 2017; Sachs et al., 2019; Stiglitz et al., 2009). Similarly, in contemporary theories of agency itself, scholars are expanding their conception of human flourishing and psychosocial well-being to accommodate richer, alternative commitments (Seligman et al., 2013). Some psychologists are exploring new theories of virtue, referencing classic thinking about holistic well-being (Fowers et al., 2021). In addition, as noted earlier, agentic hybridity is increasingly recognized in many fields. The future science of generative commitments can build on these contributions. In fact, this prospective enquiry harks back to Aristotle's (1980) concept of eudaimonia, about living a good life with practical wisdom. From this perspective, a science of generative commitments will be a science of eudaimonics. It will study how to compose and live a complete, flourishing life in a digitalized world. The inquiry would encompass all value commitments, complementing the existing study of specific types of value in economics, ethics, and aesthetics (see Di Fabio & Palazzeschi, 2015).

For example, consider digitally augmented health care. In these contexts, emerging priorities are overall health, well-being, and quality of life, or in other words, eudaimonic concerns. Systems will be designed and evaluated based on holistic human outcomes, and not simply on crude metrics of service delivery. Once maximized in this way, health care will be value based, personal, precise, and fully integrated into social life. Relevant technologies will include wearable and implantable devices. Importantly, this kind of system will require generative commitments, developing and adapting values and goals for both individuals and collectives. The responsible, augmented agents will recognize and/or generate a range of cultural, social, and personal commitments and preferences. There will also be empathic artificial agents enabling performative action generation in real time. Generative commitments will thus guide the design and delivery of services, while network management agents will supervise and govern the entire system. In this fashion, augmented agents in health care will generate commitments.

As Aristotle further understood, shared commitments underpin the good governance of communal life (Nussbaum, 2000). In ancient Athens, this was centered in the polis, its rituals, and celebrated by dramatic chorus. Whereas, in the modern period, good governance calls for reasoned public debate, the fair determination of collective choice, and participatory decision-making. In the period of digitalization, the governance of collective agency will be transformed as well. For example, civic participation could become more inclusive and globally integrated. As in the past, therefore, a new period of agentic experience will require a fresh approach toward collective governance and politics. And if history is any guide, we should expect to see more strife and struggle in this regard, as the impact of digitalization continues to grow. Institutional systems of power and influence are never easy to change, and digitalized societies will be no different. We see evidence of such conflict already, as opposing political and cultural groups struggle to control online social networks.

Science and History

Throughout this book, history is a guide. The argument consistently refers to three major periods of civilized humanity and agency: premodernity, modernity, and the contemporary period of digitalization. Among other key features, each period is characterized by stages of technological assistance: from the primitive technologies of premodernity, to the mechanical and analogue technologies of modernity, to the digital and neural technologies of the contemporary period. Hence, my argument also speaks to the history of human science, conceived as the study of human self-understanding over time. In fact, the agentic metamodels presented in this book constitute a broad framework for reconceiving the history of human science.

Some historians take an equally broad perspective on the past. This is true of the Annals School, founded in the mid-twentieth century by Lucien Febvre and Marc Bloch (2014). For them, history is a long narrative of unfolding worlds of lived experience and mentality. Politics and princes are then expressions of their periods, not the primary forces of history. Thomas Kuhn's (1970) work on paradigms of scientific investigation and knowledge is equally broad and long term. In fact, his analysis of paradigms could be restated in terms of metamodels and their hyperparameters. Each successive paradigm exhibits major shifts in core ontology, epistemology, and mechanisms of knowledge generation and diffusion. In many ways, Kuhn's view of the past aligns with the historical perspective of this book. Both identify long periods and general frameworks, although my argument articulates alternative processes, mechanisms, and metamodels and tries to bring fresh clarity and organization to this narrative.

More practically, digitalization accelerates historical time. During premodernity, rates of change were slow and often imperceptible. Societies were relatively stable and evolved slowly. For this reason, the premodern concept of historical time was expressed in legend and myth, rather than narratives of social and political change. Then during modernity, history accelerated. Indeed, for modern societies and persons, historical change is a central feature of communal and autobiographical narrative, not some distant horizon or myth, although for many indigenous cultures, the modern compression of time has been a source of cultural distress and alienation. They struggle to maintain traditional narratives in the face of imperialistic and industrial forces. In a digitalized world, history accelerates yet again. Now all humanity will share the indigenous struggle to maintain cultural narratives. In these respects, augmented humanity can look to indigenous peoples for lessons about cultural survival in the face of overwhelming social and technological change (Hogan & Singh, 2018).

For without doubt, in a digitalized world, dynamic change will be constant and ubiquitous. This will not be the end of history, by any means, but it does imply significant acceleration. Historical transformation will occur within generations and seasons, not only across the life span. Viewed positively, this will enable a new type of self-generativity, empowering augmented humanity to make and remake the world, while living within it (Latour, 2013). Augmented humanity will move "off the edge of history," as Anthony Giddens (2015) puts it, by compressing and transcending the classic parameters of historical time. Change will be discontinuous and the past will explain less and less about the future. But what comes next is not yet assured. Moving off the edge of history can be perilous or liberating. Regarding peril, some people and communities might lose their bearings or surrender to artificial control. In terms of liberation, a new period of self-generative freedom and flourishing is possible, assuming humanity meets the supervisory challenge of digital augmentation.

Research Methods

Additional consequences follow for research methods. In standard approaches, researchers in the human sciences gather qualitative data to support descriptive, interpretive models of human experience and behavior, and quantitative data to support calculative, causal models of such phenomena. The former methods focus on rich, holistic description, narratives, and sense-making, hoping to interpret meaning and intention, while the latter methods seek measurable data and discrete mechanisms, to explain causation. Multiple and mixed methods blend these approaches. Scholars debate which approach is more reliable and enlightening: rich, holistic descriptions of experience and the interpretation of meaning, or measurable mechanisms of variation in causal explanation; or some combination of these approaches (Creswell, 2003). Of course, this takes us back to Herbert Simon (1979) again, and the dilemmas of simplification in the modeling of human thought and behavior.

In parallel, scholars debate ontological and epistemological priorities. On the one hand, those who privilege qualitative methods and interpretation, typically argue that holistic description, consciousness, and meaning take priority and cannot be reduced to mechanistic cause, while on the other hand those who privilege quantitative methods and causation argue that functional mechanisms and assisted observation take priority and reject any reliance on subjective meaning and interpretation. Not surprisingly, many regard qualitative and quantitative methods as deeply incommensurable. That said, a significant research community now advocates for blended, mixed, and multiple methods (Denzin, 2010).

In a period of digitalization, these distinctions will blur even further and faster. For example, it is already possible to achieve machine-based outcomes which were previously deemed impossible, such as automated pattern recognition, associative computation, artificial empathy, intuition, and creativity (Choudhury et al., 2020; Varshney et al., 2015). Quite simply, digital systems are replicating many of the more complex, holistic functions of human cognition. As noted in earlier chapters, within the foreseeable future, there will be no detectable difference between human and artificial agents in these domains, although whether artificial agents should be classified as truly sentient and conscious is another question. Nevertheless, in consequence of these developments, it is feasible to compose blended research methods at massive scale. Different tools and techniques will be combined and recombined to match problem contexts. Studies will apply quantitative techniques to the interpretation of meaning, including self-narratives and sense-making, while also scaling qualitative techniques to predict complex patterns of thought and behavior.

Using such compositive methods, augmented science will customize different techniques of sampling and search to the phenomena and questions of interest. Here again, entrogenous mediators will play a central role, updating metamodels and methods in real time. Therefore, just as the metamodels of augmented science will be contextual and generative, so will be the methods used to gather, interpret, and analyze information. Methods will be composed to fit the problem space. They will be compositive, as Hayek (1952) originally proposed, not simply qualitative, quantitative, or mixed, in the traditional sense. In fact, advanced forms of artificial intelligence and machine learning already do this (e.g., Mehta et al., 2019). Some social scientists do as well (e.g., Latour, 2011).

Future Prospect

James March (2006), the great scholar of organizations, argues that it requires courage and positive deviance to embrace the ambiguity and ambivalence of exploratory thought. It also requires patience and persistence, to see whether fruits ripen or not. And it should, given the need for rigor and replication. However, the science of augmented agency calls for extra effort and speed, in these respects. Digitalization is rapidly infusing agentic domains, bringing unprecedented gains in capability and potentiality. In consequence, it problematizes the traditional assumptions of modernity, and presents new and urgent challenges. Most particularly, the combinatorics of digitally augmented humanity are transforming and confronting. Human agents will likely remain relatively myopic, sluggish, layered, and insensitive to variance, while artificial agents will be increasingly farsighted, fast, compressed, and hypersensitive. As both collaborate more closely, they risk amplifying the tendencies of the other, leading to internal divergence or convergence and dysfunction.

Novel problematics and dilemmas emerge. Inadequate supervision of these could produce the following dysfunctions: highly ambimodal systems, resulting in incoherent agentic form and function; highly ambiopic problem-solving and cognitive empathicing will skew judgments of the world and other minds; highly ambiactive self-regulation, evaluation of performance, and learning, would risk incoherence, extreme ambiguity and ambivalence; all contributing to dysfunctional patterns of ambidextrous, human and artificial self-generation. To mitigate these risks, human and artificial agents must develop the capability for collaborative supervision grounded in mutual understanding, trust, and respect. Achieving all this will be contingent on the development of a science of augmented agency. The core features of this science will include the following:

digitally augmented mind will be treated as a fundamental category of reality; the science of augmented agency will employ contextually sensitive, compositive methods; its metamodels will be highly generative, rather than replicative or slowly adaptive; problem sampling and representation will be intelligent and reasoned, complementing ecological rationality; augmented agency will rely deeply on the entrogenous mediation of intelligent sensory perception, performative action generation, and contextual learning; ordinary consciousness and commitments will play important roles in humanizing the science of augmented agency, rather than being sources of fundamental insight about the world itself.

Granted, the exact shape of this future science is not yet clear. Much of the current chapter—indeed, this book as a whole—is therefore prospective. It anticipates the future, grounded in the best knowledge currently available, while acknowledging that its proposals will require further elaboration and testing. Nor is this book a comprehensive treatment of the phenomena. Rather, it takes steps toward a science of augmented agency. But the process remains emergent. The trajectory of digital augmentation could change, as the natural, human, and virtual worlds continue evolving, interacting, and often conflicting. That said, we need to move forward. Prospective theorizing helps, by shedding light on unfamiliar territory. The history of science also teaches that radically new phenomena often require fresh conceptual architecture. Existing frameworks rarely suffice and waiting for certainty and normality is unlikely to succeed. Digitalization is too novel and dynamic. We could wait in vain, while the world moves on. This would be unproductive and arguably negligent, given the accelerating impact of digitalization. The augmentation of humanity has clearly begun. Its dilemmas are present and increasingly urgent. Science must respond with matching speed and purpose.

References

Appiah, K. A. (2010). *Cosmopolitanism: Ethics in a world of strangers*. WW Norton & Company.

Aristotle. (1980). *The Nicomachean Ethics* (D. Ross, Trans.). Oxford University Press.

Bandura, A. (2012). On the functional properties of perceived self-efficacy revisited. *Journal of Management, 38*(1), 9–44.

Battilana, J., Sengul, M., Pache, A.-C., & Model, J. (2015). Harnessing productive tensions in hybrid organizations: The case of work integration social enterprises. *Academy of Management Journal, 58*(6), 1658–1685.

Bloch, M. (2014). *Feudal society*. Routledge.

Buetow, S. A. (2020). Psychological preconditions for flourishing through ultrabilitation: a descriptive framework. *Disability and rehabilitation, 42*(11), 1503–1510.

Carceroni, R. L., Sanketi, P. R., Shah, S., Ozkan, D., Mariooryad, S., Tarzjani, S. M. S., Lider, B., & Ludwig, P. W. (2017). *Touchless user interface navigation using gestures*. U.S. PTO.

Chen, S., Yu, P., & Houk, K. N. (2018). Ambimodal dipolar/diels-aldercycloaddition transition states involving proton transfers. *Journal of the American Chemical Society, 140*(51), 18124–18131.

Chimirri, N. A., & Schraube, E. (2019). Rethinking psychology of technology for future society: Exploring subjectivity from within more-than-human everyday life. In *Psychological studies of science and technology* (pp. 49–76). Springer International Publishing.

Choudhury, P., Allen, R. T., & Endres, M. G. (2020). Machine learning for pattern discovery in management research. *Strategic Management Journal, 42*(1), 30–57.

Creswell, J. W. (2003). *Research design: Qualitative, quantitative, and mixed methods approaches* (2nd ed.). Sage.

Denzin, N. K. (2010). Moments, mixed methods, and paradigm dialogs. *Qualitative Inquiry, 16*(6), 419–427.

Di Fabio, A., & Palazzeschi, L. (2015). Hedonic and eudaimonic well-being: The role of resilience beyond fluid intelligence and personality traits. *Frontiers in Psychology, 6*, 1367.

Eden, A. H., Moor, J. H., Søraker, J. H., & Steinhart, E. (2015). *Singularity hypotheses*. Springer.

Fiedler, K. (2017). What constitutes strong psychological science? The (neglected) role of diagnosticity and a priori theorizing. *Perspectives on Psychological Science, 12*(1), 46–61.

Fiedler, K., & Juslin, P. (Eds.). (2006). *Information sampling and adaptive cognition*. Cambridge University Press.

Fowers, B. J., Carroll, J. S., Leonhardt, N. D., & Cokelet, B. (2021). The emerging science of virtue. *Perspectives on Psychological Science, 16*(1), 118–147.

Giddens, A. (2013). *The consequences of modernity*. Wiley.

Giddens, A. (2015). Off the edge of history. *The world in the 21st century. Teacher Training Agency Annual Lecture*, London.

Gigerenzer, G., & Gaissmaier, W. (2011). Heuristic decision making. *Annual Review of Psychology, 62*, 451–482.

Hayek, F. A. (1952). *The counter-revolution of science: Studies on the abuse of reason*. The Free Press.

Hogan, T., & Singh, P. (2018). Modes of indigenous modernity: Identities, stories, pathways. *Thesis Eleven, 145*(1), 3–9.

Hwang, H., & Colyvas, J. A. (2021). Constructed actors and constitutive institutions for a contemporary world. *Academy of Management Review, 46*(1), 214–219.

Kitayama, S., & Salvador, C. E. (2017). Culture embrained: Going beyond the nature-nurture dichotomy. *Perspectives on Psychological Science, 12*(5), 841–854.

Kuhn, T. S. (1970). *The structure of scientific revolutions* (2nd ed.). University of Chicago Press.

Latour, B. (2010). An attempt at a "compositionist manifesto". *New Literary History, 41*(3), 471–490.

Latour, B. (2011). From multiculturalism to multinaturalism: What rules of method for the new socio-scientific experiments? *Nature and Culture, 6*(1), 1–17.

Latour, B. (2013). *An inquiry into modes of existence* (C. Porter, Trans.). Harvard University Press.

Latour, B. (2017). Anthropology at the time of the anthropocene: A personal view of what is to be studied. In *The anthropology of sustainability* (pp. 35–49). Springer.

Levinthal, D. A. (2021). *Evolutionary processes and organizational adaptation: A mendelian perspective on strategic management*. Oxford University Press.

Lovelock, J. (2019). *Novacene: The coming age of hyperintelligence*. MIT Press.

March, J. G. (2006). Rationality, foolishness, and adaptive intelligence. *Strategic Management Journal, 27*(3), 201–214.

Mehta, Y., Majumder, N., Gelbukh, A., & Cambria, E. (2019). Recent trends in deep learning based personality detection. *Artificial Intelligence Review*, 1–27.

Metcalfe, J., & Schwartz, B. L. (2016). *The ghost in the machine: Self-reflective consciousness and the neuroscience of metacognition*. Oxford University Press.

Nussbaum, M. C. (2000). Aristotle, politics, and human capabilities: A response to Antony, Arneson, Charlesworth, and Mulgan. *Ethics, 111*(1), 102–140.

Ocasio, W. (2012). Attention to attention. *Organization Science, 22*(5), 1286–1296.

Pappa, G. L., Ochoa, G., Hyde, M. R., Freitas, A. A., Woodward, J., & Swan, J. (2014). Contrasting meta-learning and hyper-heuristic research: The role of evolutionary algorithms. *Genetic Programming and Evolvable Machines, 15*(1), 3–35.

Pentland, B. T., Feldman, M. S., Becker, M. C., & Liu, P. (2012). Dynamics of organizational routines: A generative model. *Journal of Management Studies, 49*(8), 1484–1508.

Quine, W. V. (1995). Naturalism; or, living within one's means. *Dialectica, 49*(2–4), 251–263.

Raworth, K. (2017). *Doughnut economics: Seven ways to think like a 21st-century economist.* Chelsea Green Publishing.

Remeseiro, B., Barreira, N., Sanchez-Brea, L., Ramos, L., & Mosquera, A. (2018). Machine learning applied to optometry data. In *Advances in biomedical informatics* (pp. 123–160). Springer.

Sachs, J. D., Schmidt-Traub, G., Mazzucato, M., Messner, D., Nakicenovic, N., & Rockström, J. (2019). Six transformations to achieve the sustainable development goals. *Nature Sustainability, 2*(9), 805–814.

Seibel, W. (2015). Studying hybrids: Sectors and mechanisms. *Organization Studies, 36*(6), 697–712.

Seligman, M. E. P., Railton, P., Baumeister, R. F., & Sripada, C. (2013). Navigating into the future or driven by the past. *Perspectives on Psychological Science, 8*(2), 119–141.

Sen, A. (1985). Goals, commitment, and identity. *Journal of Law, Economics & Organization, 1*(2), 341–355.

Shwartz-Ziv, R., & Tishby, N. (2017). Opening the black box of deep neural networks via information. *arXiv preprint arXiv:1703.00810.*

Simon, H. A. (1955). A behavioral model of rational choice. *Quarterly Journal of Economics, 69*(1), 99–118.

Simon, H. A. (1979). *Models of thought* (Vol. 352). Yale University Press.

Simon, H. (2000). Keynote address. Earthware Symposium, Carnegie Mellon University.

Skelcher, C., & Smith, S. R. (2015). Theorizing hybridity: Institutional logics, complex organizations, and actor identities: The case of nonprofits. *Public Administration, 93*(2), 433–448.

Smolarz-Dudarewicz, J., Poborc-Godlewska, J., & Lesnik, H. (1980). Comparative evaluation of the usefulness of the methods of studying binocular vision for purposes of vocational guidance. *Medycyna Pracy, 31*(2), 109–114.

Sohn, E. (2019). Decoding the neuroscience of consciousness. *Nature, 571*(7766), S2–S5.

Stiglitz, J. E., Sen, A., & Fitoussi, J.-P. (2009). *Report of the commission on the measurement of economic performance and social progress.*

Tauber, A. I. (2013). *Requiem for the ego: Freud and the origins of postmodernism.* Stanford University Press.

Tunyi, A. A., Ntim, C. G., & Danbolt, J. (2019). Decoupling managementinefficiency: Myopia, hyperopia and takeover likelihood. *International Review of Financial Analysis, 62*, 1–20.

Varshney, L. R., Wang, J., & Varshney, K. R. (2015). Associative algorithms for computational creativity. *The Journal of Creative Behavior, 50*(3), 211–223.

Wang, B., Xu, S., Yu, X., & Li, P. (2015). Time series forecasting based on cloud process neural network. *International Journal of Computational Intelligence Systems, 8*(5), 992–1003.

Wilson, T. D., & Dunn, E. W. (2004). Self-knowledge: Its limits, value, and potential for improvement. *Annual Review of Psychology, 55*, 493–518.

Yang, Z., Dong, X., Yu, Y., Yu, P., Li, Y., Jamieson, C., & Houk, K. N. (2018). Relationships between Product Ratios in Ambimodal Pericyclic Reactions and Bond Lengths in Transition Structures. *Journal of the American Chemical Society, 140*(8), 3061–3067.

Zukowski, M. M. (2012). *Genetics and biotechnology of bacilli.* Elsevier.

Glossary of Defined Terms

Ambiactive Processes and systems which simultaneously stimulate and suppress rates and sensitivities.

Ambidextrous Able to use both hands equally well, that is, equally capable to perform complementary activities.

Ambimodal Processes and systems which produce multiple outcome states, including different types of agentic modality.

Ambiopic Processes and systems which combine nearsighted myopia and far-sighted hyperopia.

Augmented Significantly assisted by technologies and especially digital technologies.

Compositive Methods or models which are composed and customized to suit different contexts and problems.

Digitalization The transformation of goal-directed processes through the application of digital technologies.

Discontinuous Processes and systems with different levels which have significant gaps between them.

Dyssynchronous Processes and systems which cycle at different rates and are poorly synchronized.

Entrogenous Mediators of in-betweenness, whereby systems develop and transform boundaries.

Hyperactive Abnormally or extremely active.

© The Author(s) 2021
P. T. Bryant, *Augmented Humanity*, https://doi.org/10.1007/978-3-030-76445-6

Hyperheuristic Simplified means of selecting sets of related models or metamodels.

Hyperopic Farsighted expansive processes, especially in sampling and search, the opposite of myopia.

Hyperparameter Fundamental characteristics or attributes of metamodels, including their broad categories, relations, and mechanisms.

Hypersensitive Abnormally or extremely sensitive.

Maximize To partially rank order a set, and then choose a member of which is no worse any other members, given some evaluation criteria.

Metaheuristic Simplified means of selecting sets of related heuristics or models.

Metamodel The common features of a family or related set of models.

Optimize To fully rank order a set, and then choose the member which ranks above all others.

Index